CALGARY PUBLIC LIBRARY

NOV 2017

Green Tyranny

Green Tyranny

Exposing the Totalitarian Roots of the Climate Industrial Complex

Rupert Darwall

Encounter Books
New York London

© 2017 by Rupert Darwall

All rights reserved. No part of this publication may be reproduced, stored in a retrieval system, or transmitted, in any form or by any means, electronic, mechanical, photocopying, recording, or otherwise, without the prior written permission of Encounter Books, 900 Broadway, Suite 601, New York, New York, 10003.

First American edition published in 2017 by Encounter Books, an activity of Encounter for Culture and Education, Inc., a nonprofit, tax exempt corporation.
Encounter Books website address: www.encounterbooks.com

Manufactured in the United States and printed on acid-free paper. The paper used in this publication meets the minimum requirements of ANSI/NISO Z39.48–1992 (R 1997) (*Permanence of Paper*).

FIRST AMERICAN EDITION

LIBRARY OF CONGRESS CATALOGING-IN-PUBLICATION DATA
Names: Darwall, Rupert, author.
Title: Green tyranny : exposing the totalitarian roots of the climate industrial complex / by Rupert Darwall.
Description: New York : Encounter Books, 2017. |
Includes bibliographical references and index.
Identifiers: LCCN 2017009693 (print) | LCCN 2017036776 (ebook) |
ISBN 9781594039362 (Ebook) | ISBN 9781594039355 (hardcover : alk. paper)
Subjects: LCSH: Environmentalism—Political aspects—Europe |
Environmentalism—Political aspects—United States. | Climatic changes—
Political aspects—Europe. | Climatic changes—Political aspects—United States. |
Environmental policy—Europe. | Environmental policy—United States.
Classification: LCC GE160.E85 (ebook) | LCC GE160.E85 D37 2017 (print) |
DDC 304.2/5094—dc23
LC record available at https://lccn.loc.gov/2017009693

Dedicated to David Henderson and S. Fred Singer

Contents

Thus, with tragic clarity, was made manifest
a sacred law of life: human freedom stands
above everything. There is no end in the world
for the sake of which it is permissible to sacrifice
human freedom.

Vasily Grossman, *Everything Flows*

Preface

After publication of *The Age of Global Warming—A History* (2013), I was approached by a reader with professional knowledge of the subject going back to the mid-1980s. The gist of the conversation went, as I recall, you have written a good book, but like most English and American writers, you missed out on developments in continental Europe, especially Sweden and Germany. The more I looked at the evidence, the more I saw he was right.

The result is this book. Comparing the two, the summit turns out to be the same, but this one takes the reader up the dark side of the mountain. There are a handful of places that touch on episodes in the previous book, but the narrative in this one makes an ascent across new terrain.

Were it not for its impact on industrialized societies' reliance on hydrocarbon energy, theories of man-made climate change would principally be of limited academic interest. In fact, these theories were first politicized precisely because of the demands they make to decarbonize energy. Sweden debuted global warming as part of its war on coal when Al Gore was still at law school. It was meant to have ushered in an age of nuclear power. The reason it didn't, instead becoming an age of wind and solar, is principally because of Germany. Despite being Europe's premier industrial economy, German culture harbors an irrational, nihilistic reaction against industrialization, evident before and during the Nazi era. It disappeared after Hitler's defeat and only bubbled up again in the terrorism and antinuclear protests of the 1970s and the formation of the Green Party in 1980.

German eco-ideas found their way across the Atlantic, where they fed progressives' attack on capitalism, targeting its most concrete manifestation—the modern corporation. In the 1940s, the Austrian economist Joseph Schumpeter identified the corporation's vulnerability in his prediction that capitalism would be the cause of its own downfall. Environmentalism—the belief that mankind's activities threaten the survival of both humanity and the planet—found a receptive audience in boardrooms and among those to whom business leaders turn to tell them what is important.

Concern about global warming became hardwired into elite thinking just as America was on the cusp of becoming the world's hydrocarbon superpower. Whereas wind and solar energy comport well with the lassitude of a European continent in relative decline, the energy revolution propelled by hydraulic fracturing—fracking—reflects the continued vitality of American capitalism. For Europe's green radicals, control of energy policy is a means toward an end. Global warming thus poses a question about the nature and purpose of the state: whether its role is to effect a radical transformation of society or whether its principal task is to protect freedom.

President Trump's announcement that the United States is to withdraw from the Paris climate agreement is historic. The Paris Agreement was the closest the Europeans had come to getting the United States to accept timetabled emissions cuts in the quarter-century saga of the UN climate change talks. The first occasion was during negotiations for the 1992 UN climate change framework convention. That attempt was rebuffed by George H. W. Bush. The second was in the 1997 Kyoto Protocol, signed by the Clinton Administration, effectively vetoed by the Senate, and then repudiated by George W. Bush. In many respects Donald Trump's rejection was the cruelest of all. The agreement's entire architecture had been designed to circumvent the Constitution's requirement for the Senate's advice and consent, a compromise the Europeans reluctantly accepted in return for the certainty of American participation.

In fact, Trump offered the Europeans an olive branch in renegotiating the Paris Agreement or negotiating a new agreement. Within minutes, the leaders of Germany, France, and Italy slammed the door shut, stating that the agreement "cannot be renegotiated."[1] Progressively outlawing the use of hydrocarbon energy would offer substantial economic opportunities, the Europeans averred, before offering to step up their efforts to assist developing nations. If cutting emissions creates wealth, why the need for hundreds of billions of dollars of climate finance?

In this respect, President Trump has a surer grasp of the economic realities than the Europeans. The United States is now the world's hydrocarbon superpower. Thanks to fracking, it has surpassed Saudi

Arabia and Russia to become the world's top energy producer. This abundance of hydrocarbon energy made the United States the biggest loser from the Paris Agreement. Quitting Paris, in what hedge fund manager Tom Steyer called a "traitorous act of war against the American people," turns the United States into the biggest winner from Paris.[2] Access to cheap energy gives American businesses and workers a colossal competitive advantage in world markets as other nations increasingly burden themselves with high-cost, unreliable wind and solar energy.

What is it about wind and solar? Sanctimonious European leaders parading their moral superiority overlook Germany's epic climate fail. The fall in German power-station emissions stalled and then began to reverse as wind and solar capacity increased. Germany's *Energiewende* (Energy Transition) is reckoned to cost up to €1 trillion ($1.12 trillion) by the end of the 2030s. Yet the big falls in carbon dioxide emissions happened in the wake of German reunification (cost: €1.3 trillion) as East Germany's inefficient, communist-era economy was closed down.

What they do tells us more than what they say. If European politicians were truly motivated by concern about global warming, they would be extending the lives of their nuclear power stations and not accelerating their closure, as German chancellor Angela Merkel has done and France's president, Emmanuel Macron, is doing with the appointment of the green activist and filmmaker Nicolas Hulot as energy minister. Similarly, Democratic governors in California and New York are closing down their states' nuclear power stations even though they are the only reliable provider of genuine zero-emission electricity other than hydroelectric power.

This preference reveals something profound about the political purpose and ideological nature of the man-made climate crisis. The climate war isn't about climate; it is and always has been about energy, and from that to transform modern, industrialized societies dependent on low-cost hydrocarbon energy. In the mind of Bill McKibben, one of the leading voices of the Climate Industrial Complex, climate denial now is twinned with something as ugly and insidious: "renewable denial," demonstrating that the aim of the Climate Industrial Complex is the replacement of hydrocarbon energy with wind and solar.[3] McKibben's antihydrocarbon ideology can't hide a crucial fact. In the thirteen years that Germany ramped up wind and solar capacity only to see its power emissions rise, American power station emissions fell ten times more than German power emissions rose.

"The voice of the people remains the most powerful force in the land," Steyer tweeted after Trump announced his decision on the Paris Agreement.[4] Steyer's is not the voice of the American people. He happens to be the most vocal of the billionaire funders of the Climate Industrial Complex, an amalgam of American money and European ideas seeking to end American exceptionalism. It is no historical accident that the United States is the only advanced country to have rejected targets and timetables in a climate change treaty, let alone three times. Delivering preordained emissions cuts requires a powerful administrative state. Uniquely, America's Constitution and its separation of powers provide checks against it. This, ultimately, is what is at stake in the battle of Paris and the climate war. It is a fight for America's soul.

Notes

1 Government of Italy, "Declaration of the Italy, Germany and France on the announcement of the USA to leave the Paris Climate accord," June 1, 2017.

2 Tom Steyer tweet, May 28, 2017, https://twitter.com/tomsteyer/status/869018773453291521 (accessed June 19, 2017).

3 Bill McKibben, "The New Battle Plan for the Planet's Climate Crisis," *Rolling Stone*, January 24, 2017, http://www.rollingstone.com/politics/features/bill-mckibbens-battle-plan-for-the-planets-climate-crisis-w462680 (accessed June 19, 2017).

4 Tom Steyer tweet, June 1, 2017, https://twitter.com/TomSteyer/status/870368866949488640 (accessed June 19, 2017).

Acknowledgments

The views and conclusions are mine, but in researching and writing this book, I'd like to express thanks to Jacob Arechiga, Stefan Björklund, John Constable, Chuck DeVore, Myron Ebell, Edgar Gärtner, Steven Groves, Kathleen Hartnett-White, Chris Horner, Gordon Hughes, Hans Lundberg, Nick Loris, Mark Mills, Philipp Müller, Michael Nasi, Benny Peiser, Jeremy Rabkin, Birgit Reinhardt, Bo Theutenberg, Fritz Vahrenholt, and Alexander Wendt. Ian Tanner read and corrected the manuscript, which was subsequently edited by Barry Varela, designed by Chris Crochetière of BW&A Books, and overseen by Katherine Wong and Sam Schneider at Encounter Books. My agent, Keith Urbahn of Javelin, tirelessly and cheerfully championed the project throughout. A generous grant from the Searle Freedom Trust made this book possible. I am grateful to them all.

An author inflicts the greatest burden on his family. This one is no different and was repaid with tolerance and understanding by my wife. From the bottom of my heart, thank you.

The combination of intelligence and integrity is far rarer than it should be, especially in a field such as this. S. Fred Singer and David Henderson have both in abundance. As well as providing invaluable insights, Fred Singer features as a protagonist in two of the episodes recounted in the pages of this book, a role for which he has been traduced and attacked to this day. Long conversations with David Henderson inspired and shaped my first book and made this one much better. We are now collaborating on a third, which is on a different subject. Both men are models of what an intellectual warrior dedicated to truth seeking should be.

This book is dedicated to them.

1 America in Lilliput

I attempted to rise, but was not able to stir:
for, as I happened to lie on my back, I found my
arms and legs were strongly fastened on each
side to the ground; and my hair, which was long
and thick, tied down in the same manner.

Jonathan Swift, Gulliver's Travels

If only everyone could be like the Scandinavians,
this would all be easy.

Barack Obama[1]

This book is about freedom. It is about its loss as a result of policies designed to slow down what is presumed to be man-made global warming. Avoiding planetary catastrophe gives a president and the executive branch a higher dispensation than that granted by the Constitution. Obamacare was implemented under the Affordable Care Act. Implementation of the Clean Power Plan was by administrative fiat and the Senate bypassed when the United States ratified the 2015 Paris Agreement: America's eighteenth-century Constitution is not going to be allowed to impede a project in which society is to be radically transformed through the agency of the state. As the embodiment of an ideal of freedom, the Constitution is incompatible with a project that is alien to the tradition of liberty flowing from America's founding, though not to the ideologies, originating in Europe, from which the project first sprang. The two cannot coexist. One or other will prevail and define America for decades to come.

The vast gap between American hard power and that of the rest of the world sometimes blinds Americans—especially American conservatives—to America's vulnerability to other countries' soft power. America invented Earth Day in 1970 and gave birth to postwar environmentalism with Rachel Carson's *Silent Spring* (1962). Yet even these seemingly all-American products drew on ideas from across the Atlantic and from across the chasm of the Second World War; the cancer chapter in Rachel Carson's *Silent Spring*, for instance, incorporated the Nazi belief that industrialization was causing a cancer epidemic.

If there was a purely American strand of environmentalism, the demands it made on America were fairly limited. The costs of banning DDT—the principal policy consequence of *Silent Spring*—were mainly inflicted on Africans exposed to the risk of malaria. Thanks to the availability of cheap substitutes, phasing out CFCs a decade and a half later to preserve the ozone layer hardly required Americans to change their lifestyles. Preserving habitats and wildernesses did not necessitate transforming American society and culture.

There is a strand of American apocalyptic thinking that was first initiated by scientists after 1945 in reaction to the atomic bomb. But this scarcely amounted to an ideological challenge to the basis of American capitalism. That came when it was mixed with the post-Marxist environmentalism developed by German exiles and subsequently weaponized by the progressive left in America. They were the prophets who prepared the way. Their student followers in Germany would come to form the leadership of the Greens in the early 1980s. Nazi ecological politics were rehabilitated by the Greens and would come to form part of mainstream German and then European politics. What united them was a deep hostility to capitalism and the free market. Against them stands the Jeremiah of capitalism. Far from wishing to see capitalism fail, Joseph Schumpeter foresaw its death coming from its own hand; although writing in the 1940s, he could not have foreseen that the instrument of its self-destruction would be environmentalism.

This, then, is the ideological landscape across which the action unfolds. At the end of the 1960s, while American environmentalists were focusing their efforts on banning DDT, Sweden was putting coal—the most ubiquitous source of electrical energy—in the crosshairs when it made acid rain the world's top environmental problem. By making energy the focus of international action, it gave environmentalism control of the dial to transform the basis of industrial civilization.

In the past, waves of spontaneous innovation transformed the fabric of American society, vastly improving Americans' quality of life. None was as transformational as cheap, ubiquitous electrical power. It bade farewell to the age of steam, gaslight, and paraffin. Grid-supplied electricity separated the twentieth century from the nineteenth and triggered a social revolution. Electrical appliances replaced domestic servants, and the liberation of women from household drudgery began.

This one is different—a planned societal and cultural transformation directed by the administrative state. Within a year of Barack Obama's election to the White House, such a transformation was being discussed by European climate change radicals and senior Democrats at a conference in Germany. Where Europe led, America would follow.

This book tells the story of two countries and three environmental scares. Two originated in Sweden (acid rain and global warming) and one (the nuclear winter) was transmitted from Moscow via Stockholm.

Sweden

Sweden was to have an influence on world affairs and the environmental politics of the United States out of all proportion to its eight million people. From the late 1960s through to the late 1980s and the establishment of the Intergovernmental Panel on Climate Change (IPCC), which, far more than any other nation, has the rightful claim to paternity (Chapter 11), Sweden was extraordinarily successful at projecting environmental diplomacy on the world stage. Its Social Democratic rulers, in particular its prime minister, Olof Palme, had compelling reasons to mobilize environmentalism for political ends. A widely unpopular civil nuclear power program would, in 1976, help bring to an end 48 years of unbroken Social Democratic rule.

Acid rain (Scare #1) was the dress rehearsal for global warming. The politicized science of acid rain swept all before it, the bar set low in the first government report on acid rain, which happened to be written by Bert Bolin, a friend of Palme and future first chair of the IPCC (Chapter 6). It spread to Germany, where hysteria about "forest death" destroyed any hope of rationality and objectivity. It was taken up by Canada, which waged a relentless campaign to get the United States to cut its power station emissions. The Reagan Administration held firm against virtually unanimous scientific opinion. Elected as the environmental president, George H. W. Bush gave the Canadians what they wanted. However, the science was not as solid as the consensus asserted, and a ten-year federal study revealed it for what it was. Scandalously, the Environmental Protection Agency (EPA) suppressed its findings until the main provisions of the acid rain legislation had been agreed in Congress (Chapter 7). Despite its falsification, as of this writing, the EPA still maintains that sulfur dioxide from power station emissions makes lakes acidic and damages forests and woodlands.

Environmentalism became a tool in the Cold War. After regaining power in 1982, Palme proceeded to play a dangerous and duplicitous game in the Cold War's climactic decade. In being used as a drop box for the nuclear winter scare that had been concocted in Moscow (Scare #2), Sweden undermined the interests of the West in the nuclear rearmament showdown that was to end the Cold War. So did the American climate scientists and the scientific community generally, who abused their standing as scientists to peddle a scare story that had no objective rationale other than to favor the Soviet Union (Chapter 10).

Sweden's neutrality enhanced Palme's ability to project his nation's moral superiority. It had the interests of the planet at heart, not those of the Cold War blocs, and preened itself as the moral conscience of the world. It denounced American imperialism in Africa and Southeast Asia and championed Third World liberation movements, such as the genocidal Khmer Rouge, and promoted nuclear disarmament and environmentalism. But Sweden was not what it appeared to be. Its neutrality was based on a lie. From the early years of the Cold War, Sweden had a secret military alliance with Washington. Naturally, American leaders knew about it, as did the Soviets. The Swedish people did not. The Swedish state practiced a gross deception on its own people. This deception matters. If it had been the Italian state that had tried to put acid rain and global warming onto the international agenda, it would have gone nowhere. Sweden is different. It is not Italy. It is taken seriously. You might disagree with its positions, but it is seen as trustworthy and reliable, like the cars it makes. The deception over Sweden's alignment with the Atlantic Alliance shows this assumption to be false (Chapter 8).

To American progressives, Sweden is also a model of what they want America to be: technocratic, egalitarian, and politically correct—the embodiment of values antithetical to those of the American Revolution. Swedish society is the product of a conformist culture that inculcates unquestioning submission to authority, Roland Huntford wrote in his 1971 book, *The New Totalitarians*. Sweden did not feudalize, so it missed out on that phase of European history from which emerged Western civic values. The Reformation was used to nationalize religion. The Church was turned into a government department, and the clergy into ordained bureaucrats.[2] Religious conformity was enforced strictly and bloodlessly. The last Catholic convert was deported in 1855, and Swedish Baptists left in droves for America.

Nearly two centuries before Napoleon and four centuries before Barack Obama, Sweden had a centralized administrative apparatus. It gave the Social Democrats, who formed their first government in 1921, a political system

> adapted to the swift enactment of the intentions of the central bureaucracy. The legislature is weak, the executive strong, and, for centuries, real power has lain in the government administrative machine.[3]

The organs of the Swedish state became the perfect instruments to carry out the twentieth century's most prolonged and thorough experiment in social engineering outside the Soviet Union and China. It pioneered a cradle-to-grave welfare state to replace the family and operated a eugenics program championed by Swedish economists and scientists alike, including the leading climate scientist of the era, Svante Arrhenius (Chapter 3).

Renewable Energy

In 1988, the year of the Toronto conference on global warming as well as the establishment of the IPCC, renewable energy had not been touted as providing salvation from global warming. Under Palme, Sweden's Social Democrats plowed ahead with their nuclear power program. Just as they'd hoped, global warming (Scare #3) would, lead to an intensification of the war on coal that had been started with the acid rain scare. But it would not usher in the new era of nuclear power. Instead it would be an age of expensive and unreliable wind and solar power.

The case for renewables is both irrational and contradictory. On the one hand, it was argued that, unlike fossil fuels, they wouldn't run out. Despite a century and a half history of missed forecasts, even if fossil fuels were in terminal decline, there is no sense in leaving them in the ground. Their only value is the use of them we make above ground. On the other hand, mitigating global warming requires curbing promiscuous use of fossil fuels, implicitly conceding they're not going to run out after all, otherwise higher prices from growing shortages would automatically do the job.

Unlike any other form of energy or, indeed, any other commodity, electricity can't be stored (it has to be reconverted into other forms of energy) and has to be produced the moment it is consumed. Before renewables came on the scene, demand determined supply, yet the amount of electricity produced from wind and solar farms depends on the weather. Wind and solar thus suffer from a fundamental flaw: They are incapable of supplying on-demand electrical power.

Their inability to meet a basic requirement of electricity production means other generators, typically powered by fossil fuels, have to be kept on standby to pick up the slack. This is highly inefficient and adds to costs and to greenhouse gas emissions. Wind and solar also require more grid infrastructure. The more wind and solar on the grid, the worse it is for the electricity system taken as a whole. Even if it becomes cheaper to make and install wind turbines and solar panels, any plant-level economies of scale are more than outweighed by the system diseconomies of scale. Beyond the most limited applications, wind and solar energy doesn't make economic sense.

Germany

The reason why renewables took off is to be found in the politics and culture of the second country examined in these chapters. In the 1930s, when Sweden barred fleeing Jews from entering the country, the Nazis became the first political party in the world to promote wind power. Virtually all the themes of the modern environmental movement are prefigured in this, the darkest chapter of European history, making it a deeply problematic period for environmentalists (Chapter 4).

After World War II, West Germany repudiated the irrationalism and nihilism of the Nazis and became the European linchpin of the Atlantic Alliance. West Germany was governed well and democratically—and Germans prospered mightily. Though scattered, antidemocratic forces had not entirely disappeared. After his denazification, the philosopher Martin Heidegger preached anti-industrial, environmentalist metaphysics. Marxist intellectuals of the Frankfurt School returned from exile in the United States, where they had developed the New Left's synthesis of Marxism and environmentalism and spread their antidemocratic antirationalism across American universities (Chapter 5).

Back in Germany, the Frankfurt School influenced a generation of antidemocratic radical leftists. The 1968ers rejected West German democracy, claiming it the latest manifestation of fascism. Some went to Palestinian training camps and became terrorists. The hijackings, bombings, and shootings directed at Jews and capitalists culminated in the German Autumn of 1977. Having set out to fight Nazism, as the New York writer Paul Berman puts it, the Sixty-eighters ended up "Nazi-like" (Chapter 9). Defanged and marginalized, it was not the end of it. The late 1970s saw a growing antinuclear protest movement with more than a smattering of ex-Nazis and neo-Nazis. The remnants of the Sixty-eighters, led by Germany's future foreign minister, Joschka Fischer, saw their opportunity. Three years later, they formed the leadership cadre of West Germany's newest political party—the Greens. The Sixty-eighter and New Left embrace of environmentalism was indeed "Nazi-like" in reviving the green tenets of the Nazi era.

West Germany was in the frontline of Moscow's attempt to split the Atlantic Alliance and win the Cold War. The early 1980s saw massive anti–North Atlantic Treaty Organization (NATO) demonstrations sweep the country. The antinuclear movement quickly became the peace movement and was heavily penetrated by Eastern bloc intelligence services. The emergence of an environmentalist party to the left of the German Social Democrats, the SPD, was to have profound and lasting consequences on German and European politics. Within a decade, the people who had been on the wrong side of the Cold War came out on top. In 1998, Fischer became vice chancellor and foreign minister in Berlin's first Red–Green coalition.

Two years later, the German parliament passed the Renewable Energy Act. The law would unleash a renewable energy gold rush and a solar boom, though the green jobs turned out to be Chinese rather than German. The Chinese solar boom led to a collapse in the price of solar photovoltaic (PV) panels, but electricity bills kept going up. Why solar? Here again there would be resonances with Germany's pre-1945 past. The sun had an important place in Nazi symbolism (the swastika). The Greens' sunflower logo was designed by a former Nazi, Joseph Beuys. Germany's chief solar lobbyist and architect of the renewable energy law was inspired by an early-twentieth-century scientist, Wilhelm Ostwald. Much as modern environmentalists do, Ostwald believed that society's energy consumption should be no greater than what the Earth receives each year from the sun (Chapter 12).

The renewable energy transition would cost the equivalent of a scoop of ice cream on monthly electricity bills, the Greens claimed. Nine years after the scoop of ice cream came a revised price tag—€1 trillion ($1.13 trillion). Soaring electricity bills led to a consumer backlash. Renewables turned out to be an engine of financial destruction, costing shareholders of the three quoted German utility companies nearly €70 billion ($79b) and consumers €269 billion ($304b) in higher electricity bills. German chancellor Angela Merkel knew that the cost of renewables would be ruinous, but political advantage trumped rationality. To squeeze the SPD between the Greens to their left and her own Christian Democrats to their right, in 2007 she pushed the European Union to adopt a mandatory renewables target (Chapter 13).

Europe

Europeans started to learn a painful lesson: The full cost of wind and solar is a lot higher than the sticker price. But Germany's—and Europe's—push for renewables wasn't about rationality. It was a product of German culture and green ideology (Chapter 14).

To sustain the flow of government support for wind and solar, an immensely powerful Climate Industrial Complex came into being, fronted by environmental NGOs. The use of NGOs as shock troops to overwhelm business opposition to environmental protection had been envisaged by a top German government bureaucrat, Günter Hartkopf, in a 1986 address to his civil servant colleagues. Dense networks of bureaucrats, academics, environmental activists, and lobbyists enabled the Climate Industrial Complex to control the sources of advice that ministries and regulatory agencies rely on and contributed to the lack of proper scrutiny that might have prevented the largest misallocation of resources in history. No one had worked out how wind and solar was going to be integrated into the electricity system and what it would all cost. Government support for wind and solar was less about assuring the survival of the unfittest than guaranteeing the triumph of the unfittest (Chapter 15).

Power without responsibility, British prime minister Stanley Baldwin said of the press barons of the 1930s—the prerogative of the harlot throughout the ages. That role came to be played by environmental NGOs. Indispensable to the triumph of the Climate Industrial Complex, headline-generating NGOs such as Greenpeace, Friends of the Earth, and the World Wildlife Fund (WWF) had at first been wary of global warming, viewing it as a Trojan horse for the nuclear power industry, which they had spent most of the 1980s fighting. Nature conservation had been their original mission, but in the 1990s, NGOs began pimping themselves to the wind and solar industry, the price being paid by wildlife and the environment. Wind farms are especially harmful to birds and bats, yet NGOs promote wind and attack nuclear, which has minimal environmental impacts. NGO concern about nature is selective. Thus the mass eco-slaughter of raptors and migratory birds by the rotating blades of wind turbines is necessary. Wind farms, good; coal and nuclear, bad (Chapter 16).

Seven years after taking the plunge with mandatory renewables targets, their dire economic consequences led the European Commission to have second thoughts. Then there is what Germans call "Das CO_2 Paradox." Wind and solar aren't much good at cutting carbon dioxide emissions. As wind and solar capacity rose, carbon dioxide emissions barely fell. Between 1999 and 2012, German power station emissions actually rose. It is a paradox only if it is assumed that the real purpose of renewables was to cut carbon dioxide emissions. If it had been, Germany would have encouraged hydraulic fracturing—fracking—and kept its nuclear power stations going for as long as possible. Germany banned fracking and accelerated the shutdown of its nuclear power stations. Judged by what Germany actually did, global warming was a smokescreen for a radical green agenda and a massive transfer of wealth from consumers to green rent-seekers (Chapter 17).

California

Similarly, when California's legislature passed a law in 2011 requiring one third of its electricity be generated from renewables by 2020, it excluded zero-carbon nuclear and large hydropower. In the late 1960s, California's political and intellectual class became increasingly influenced by environmentalist thinkers such as the Norwegian Arne Næss, the German Fritz Schumacher, and the Frankfurt School's Herbert Marcuse. From the mid-1970s, the Golden State led the U.S. in turning green ideology in a fight against reality, with disastrous consequences. Rising demand and falling generating capacity, exacerbated by retail price caps, led to grid instability and rolling blackouts in 2000 and 2001, contributing to the 2003 recall of Governor Gray Davis. Although the blackouts ended, non–weather dependent generating capacity continued to fall, electricity prices kept rising, as did the amount of wind and solar on the grid. By 2014, California was importing one-third of its electricity. Promoting itself as a model for America, it is self-evidently impossible for all the other forty-seven contiguous states to import one-third of their electricity from each other.

California also led the way in being the crucible of America's Climate Industrial Complex. In the 1940s, the Austrian economist Joseph Schumpeter argued that cultural and sociological factors would lead to the demise of capitalism. The fruits of capitalism harvested by tech billionaires, hedge fund managers, and foundation executives were poured into the 2010 fight to defeat Proposition 23, which had attempted to limit the economic damage of California's renewable targets. With the nation's largest number of billionaires and its highest poverty rate (taking into account housing and other benefits), postindustrial California began to resemble a precapitalist society (Chapter 18).

Freedom in America

After conquering California, next was Washington, D.C. By mid-2010, Democratic leaders in Congress had given up on getting cap-and-trade legislation to the President's desk. NGOs funded by Silicon Valley oligarchs and progressive foundations colluded with the EPA in a sue-and-settle strategy to "force" the Obama Administration to regulate power plant emissions.

Preparations for the EPA's European-style energy transformation started as a genuinely American energy transformation had already been under way for the best part of a decade. Fracking was environmentalism's worst nightmare. Not only was it bringing the huge falls in power station emissions that had eluded Germany. By powering America toward energy independence, it was slaying the national security argument against fossil fuels and destroying the running-out-of-oil argument environmentalists had been using since the 1970s. Politicians wanted fracking's benefits, only their policies had nothing to do with getting them. They were a product of serendipity, of capitalism doing what it does best, yet Silicon Valley and foundation dollars were mobilized to stop it in its tracks. It was the beginning of a civil war within American capitalism (Chapter 19).

The Clean Power Plan, finalized in 2015, was the Climate Industrial Complex's single greatest triumph, but voters didn't care much about global warming and protecting the environment. Worse still, the modeling used by the Obama Administration showed that the bulk of any climate benefits flowed to other countries. Climate regulation was therefore pitched to voters as being about enhancing air quality and protecting public health—both achievable at virtually no cost. The EPA duly concocted a regulatory assessment that magicked away the costs of renewables (reminiscent of the scoop of ice cream claim made by the German Greens), created spurious health benefits (much as had happened after the collapse of the science behind the acid rain scare) and depended on the far-fetched assumption that Americans would, in defiance of economic history, start consuming less electricity (legalizing pot smoking and indoor cultivation of marijuana threatened to blow a hole through that) (Chapter 20).

The Clean Power Plan was different from what had gone before. In 1990, Congress had passed anti–acid rain legislation with large majorities, the economic consequences of which were miniscule in comparison. Two years later, the Senate had given its unanimous advice and consent to the UN Framework Convention on Climate Change after the executive branch had promised that any future protocols would be submitted to the Senate. Absent congressional authorization, justification for the Clean Power Plan and the December 2015 Paris climate agreement rested on something of a circular argument: The Clean Power Plan shouldn't be overturned because it formed part of America's international commitments under the Paris Agreement, but the Paris Agreement did not require the Senate's advice and consent because the agreement did not commit America to anything new.

Global warming raises profound questions about the purpose of government and the nature of limited government. Should attempts to avert a planetary catastrophe be prevented or obstructed by 4,400 words written on eighteenth-century parchment? Damaging as the economic consequences of renewable energy were proving for Europe, global warming spoke to the highest purpose of the European polity, a cause tailor-made for the European Union's postdemocratic, supranational administrative form of governance.

To make its claim on America, global warming in turn asserts its uniqueness—that nothing like it has ever happened before, that dealing with it requires an unprecedented response; in short, that global warming is different. In the boom years before the subprime crisis and the collapse of Lehman Brothers, financial experts would explain that the economy had changed and that the past offered no insight into the present and no guide to the future. It was a self-serving delusion. "More money has been lost because of four words than at the point of a gun. Those words are 'This time is different.'"[4]

Is this time different? Does global warming change everything, or is it a product of human history?

The United States was founded so Americans could be free and governed under a constitution of separated powers. America has much more to lose, more than its economic dynamism, indeed, the very idea of itself created at its founding. At stake is more than money. At stake is what makes America unique.

2 The Great Transformation

Whom the gods would destroy, they first make mad.

Henry Wadsworth Longfellow

This is probably the most difficult task we have ever given ourselves, which is to intentionally transform the economic development model, for the first time in human history.

Christiana Figueres, executive secretary, United Nations Framework Convention on Climate Change, February 3, 2015[1]

Prophets. Predators. Parasites?

Malarial protozoa infest the mosquito's brain to attract the insect to humans. Once the human is infested, the protozoa alter the host's odor to make it attractive to mosquitoes. The *Toxoplasma gondii* parasite is even smarter. It manipulates the host's brain function so that repulsion becomes attraction and the animal loses its fear of predators and turns host into prey to complete the parasite's life cycle. This single-cell parasite makes people act stupidly, too. People with *Toxoplasma* antibodies are more than twice as likely to be involved in traffic accidents.[2]

"No challenge—no challenge—poses a greater threat to future generations than climate change," Barack Obama trumpeted in his 2015 State of the Union address; not the breakup of the international system with Russia's dismemberment of a neighboring state; the seizure of swathes of the Middle East by fanatical psychopaths; or a Europe incapable of overcoming its economic paralysis. "The Pentagon says that climate change poses immediate risks to our national security."[3] Obama was only saying what the national security elites of the West believe. Yet a 2013 study found that since 2005, the rolling 15-year temperature has fallen back to the 1900–2012 long-term trend of a rise in temperature of around seven-tenths of one degree centigrade (1.3 degrees Fahrenheit) per century.[4] A rise of less than one degree centigrade over the course of a century would hardly portend crisis, let alone a planetary emergency.

Another facet of the global warming pathology that defies the logic of surviving and thriving is the urge to make energy—especially electrical energy—more costly and less abundant. More than any other technology, electrical power separates our age from the Victorians. The Industrial Revolution revolutionized factories. Electricity created the modern home—libraries, parlors, sewing rooms, were replaced by more open-planned styles, and houses didn't need to accommodate servants. Kitchens were smaller; electrical appliances replaced domestic servants and began to liberate women from the "servitude of housework." Rooms were lighter and cleaner. "We will make electric light so cheap that only the rich will be able to burn candles," Thomas Edison said, launching his electric light bulb.[5]

The new era began in New York City at 3 P.M. on September 3, 1882, when Edison switched on the first incandescent bulbs powered by his Pearl Street generator several blocks away. The few power plants of the early 1880s were inefficient, steam-driven dynamos housed in basements or adjacent properties. John Pierpont Morgan's Madison Street mansion had electric lights, "but the great banker was one of the few people in the world who could afford them."[6] Edison knew his system had to provide light at a lower cost than gas. "All parts of the system must be constructed with reference to all other parts," Edison said, "since, in one sense, all the parts form one machine, and the connections between the parts being electrical instead of mechanical."[7]

High-resistance light bulb filaments reduced the cost of copper wiring used for mains transmission. Edison knew the cost of distribution had to be further reduced, but a technical solution eluded him. Alternating current enabled long-distance transmission—"how do they make the current go the other direction?" Edison asked—unlocking the energy potential of Niagara Falls.[8] In 1895, Nikola Tesla and George Westinghouse succeeded in generating electricity there, and the following year the Niagara Falls Power Company began transmitting power to Buffalo. To this day, hydroelectric power is still the only renewable technology that has been efficiently integrated at scale.

The more houses that were connected, the more electricity they used through the day (and night), the faster the cost per kilowatt hour (kWh) could fall. In the United States, sales of electrical appliances nearly quadrupled from $23 million in 1915 to $83 million in 1920. In 1912, only 16 percent of American homes had electric lighting. By the end of the 1920s, nearly every home did.[9] Edison produced electric kettles and irons. General Electric, formed in 1892 from the merger of the Edison General Electric Company and a rival, then brought to market a stream of electrical appliances, beginning with electric fans (1902), toasters (1905), the Hotpoint electric range (1910), and refrigerators (1917). During the course of the twentieth century, electricity got cheaper and the world electrified.

Politics is now turning the page on that era.

On a rainy day in June 2009, some of the world's most prominent and influential climate thinkers assembled in Essen, Germany. Unlike climate conferences in glamorous cities or tropical islands, Essen is in the heart of the heavily industrial Ruhr coalfield. Few people go there unless they are on serious business. "The Great Transformation—Climate Change as Cultural Change" was very serious—to end the era of energy abundance initiated by Edison 126 years earlier.

The Essen conference attracted little publicity. There were no big announcements and no big decisions. There was only a smattering of environmental NGOs and a few journalists, but no television or overseas media. The conference was not about public relations (PR). It was about something much more important—how to fundamentally change Western society and the democratic values that have come to define it. Above all, the conference would demonstrate two things. It highlighted global warming's radical, postdemocratic agenda and how Europe was leading this project, with America following. In doing so, the conference raised the most fundamental question of all: Are climate policies compatible with the survival of democracy?

The 450 or so delegates included council members of Germany's powerful Advisory Council on Global Change, known by its initials WBGU, appointed and funded by the German government. Founded in 1992 in the run-up to the Rio Earth Summit, the WBGU is charged with raising public awareness and heightening the media profile of "global change issues."[10] They were joined by colleagues from the Essen Institute for Advanced Study in the Humanities (KWI). With a strongly Marxist pedigree, the KWI's leading lights include long-standing members of the German Communist Party, the DKP. Small in votes at election time, the party has considerable influence in universities and the German media. Edgar Gärtner, a former Marxist and now a free-market climate skeptic, bumped into many of his former DKP comrades during the two and a half days of the conference.[11] Alongside them was the managing director of Washington, D.C.'s Brookings Institution, ranked the most influential think tank in the world. Brookings's head, William Antholis, was in Essen to deliver his view of the good, the bad, and the ugly of climate politics. Antholis would spare nothing when it came to describing the bad and the ugly of America's democratic institutions. Banks and financial institutions, including Swiss Re, one of the world's top two reinsurers, with $232.7 billion in assets, were also present.[12] The conference's big draw was John Podesta, Bill Clinton's last White House chief of staff, head of the Obama presidential transition, and future senior counselor to President Barack Obama before becoming chair of Hillary Clinton's 2016 presidential campaign.

Hans Joachim Schellnhuber, co-chair of the WBGU and director of the Potsdam Institute for Climate Impact Research, introduced the conference. For many years, Schellnhuber was Angela Merkel's top climate adviser and scientific adviser to the president of the European Commission. He has compared himself to Albert Einstein, but Schellnhuber's math can lead him astray. The IPCC's bogus claim that the Himalayan glaciers would melt by 2035 was correct and "very easy to calculate," he told an interviewer in 2009.[13] Schellnhuber's extreme views arouse little controversy in Germany. A member of the Club of Rome, Schellnhuber advocates 1970s-style limits-to-growth politics updated for the age of global warming. Limiting global warming to two degrees centigrade above preindustrial levels should be written into international law, Schellnhuber argues. Countries should have the right to file climate protection lawsuits, something that came a step closer in 2013 when negotiators established a loss-and-damages mechanism at the Paris climate conference.[14] An elite of scientific guardians should be embedded in the legislative process so that world climate policy is kept on course. "Without a compass, you cannot steer a ship," Schellnhuber says.[15] A new constitutional settlement should enable judges to overturn the will of the majority, something one of Schellnhuber's rare German critics characterized as the legitimization of a soft dictatorship.[16] Schellnhuber's own moral compass points to the duty of scientists to safeguard the world. He draws an implicit moral equivalence between the United States and Germany in the Second World War. With events like Auschwitz and Hiroshima, history has shown you cannot expect that everything will be all right, he once said in accepting one of his many awards. "Once you have tried the apple of knowledge and are cast out of the paradise of ignorance," Schellnhuber said, "you bear a great responsibility."[17]

"The solutions to our climate-change crisis will ultimately reside largely at the level of culture," Canadian political scientist Thomas Homer-Dixon told delegates in the conference's first session. It requires questioning cultural assumptions that committed Western societies to economic growth and going beyond what Homer-Dixon called "purely procedural democracy."[18] Democracy came in for a lot of criticism the next day. Harald Welzer, a social psychologist at KWI whom Gärtner knew well from his days in the Kremlin-backed DKP, spoke of the need for Capitalism 3.0 to ensure the primacy of politics over the market and avoid the worst ecological consequences of global warming. A cultural revolution was needed to transform capitalism.[19] And it needed something extra—to change perceptions of reality and a change in moral values. This had been done before, Welzer said, under the Third Reich during the years 1933 to 1941. Welzer's theme was taken up by the next speaker, Andreas Ernst, an environmental psychologist, who told delegates that the "architecture of the human brain" made it hard for humans to do the right thing.[20]

Claus Leggewie, coauthor with Welzer of the 2009 book *The End of the World as We Know It: Climate, the Future and the Prospects for Democracy*, talked of a diminishing interest in democracy. Democracy was simply "too slow."[21] British political scientist David Held of the London School of Economics, adviser to Muammar Gaddafi's son Saif al-Islam on his fake Ph.D. thesis and director of a research program funded by Gaddafi's charity, suggested democracy was on trial.[22] "The countries that have the best records in reducing carbon emissions are all democracies," Held argued. But if they failed to successfully address the challenge of climate change, it could have "deep and profound consequences, both for what people make of modern democratic politics and the idea of rule-governed international politics."[23]

Brookings's William Antholis dispensed with even a qualified endorsement of democracy but plowed straight into his analysis of why the European political system is good for the climate and America's democracy is bad. "The EU's moral mission and their founding spirit has been [the] driving force in building a global climate regime," Antholis stated.[24] He gave short shrift to the world's most durable written constitution as a "short two hundred years" of constitutional history to contrast America's Constitution unfavorably with Europe's "two millennia of experience of constitutional construction," a curious way of characterizing the history of a continent from the time of the Roman emperor Caesar Augustus. Since George Washington became President, France has had five republics, two empires, two monarchies, and a military regime (not counting the Reign of Terror, a directory, and a consulate), and since 1871, when Germany was first unified, it has had two Reichs and two republics, has started two world wars, and was divided between east and west for 41 years.

Europe became a climate leader by empowering minority political parties, Antholis argued. By winning 10 percent of the vote, the Greens got to be 20 percent of the governing coalition. "In the United States, ten percent of the vote would be a laughable also-ran. In Europe, ten percent is a mandate to change the world." Europe's private citizens, NGOs, and corporations had also "moved the needle." Europe viewed national sovereignty as a problem and now has a "model for global governance," whereas the U.S. Senate guards "US sovereignty with all the religious fervor that was associated with that word when the Thirty Year's [*sic*] War gripped this continent six hundred years ago.'* A deep challenge lay within the American political system, which Antholis simply characterized as "bad." Its federal system—particularly the Senate—empowers minorities to block action, though Antholis forbore from reminding his audience that the Senate had unanimously given the Kyoto Protocol the thumbs down, instead blaming the Bush Administration for America's "ugly" declaration of

* Most historians date the establishment of state sovereignty as the basis of the international system to the Peace of Westphalia (1648), which ended the Thirty Years' War rather than causing it.

independence from Kyoto. American politics prioritize economic performance almost entirely to the exclusion of other policy priorities, he complained. To get around the Senate and enable the United States to be a credible participant at the Copenhagen climate summit, the plan was to create facts on the ground: first pass the Waxman-Markey cap-and-trade bill, then base the international negotiations on that law.

In what Gärtner says was one the most compelling presentations of the conference, John Podesta put more flesh on the bones of this strategy.[25] "I can say with great confidence that the new American president sees climate change as the challenge of our time," he told them, emphasizing the point with the spurious claim that 300,000 deaths a year are already attributable to climate change.[26] "Without a doubt, change has arrived," as he castigated the Bush Administration for deliberately blocking progress on the climate change negotiations.* "The United States is ready, willing, and able to negotiate an aggressive climate treaty in Copenhagen," he declared. Congress would pass "comprehensive climate legislation" within the next eighteen months. Energy transformation was not only the centerpiece of President Obama's environmental policy, but also his economic program to produce growth and jobs. The new administration had canceled oil drilling leases on 130,000 acres of protected land and was spending $71 billion on clean energy with a further $20 billion in loan guarantees and tax breaks, equivalent to around $800 per American household. Global cooperation and commitment were the only paths to a more sustainable future and to save fragile ecosystems from climate-related disasters. "We need a profound change in our culture and ethics," Podesta concluded. "We must re-evaluate our old assumptions about what is possible and I appreciate that conferences like these assist with that process." Not mentioned was America's shale revolution, which would be the one bright spot of the Obama economy for growth and jobs and which would cut America's greenhouse gas emissions by more than any other nation's.

In a panel discussion, Podesta shed more light on the administration's strategy of legislate first, treaty later. By 2020, America would have 100

* In reality, the aims of the Obama and Bush administrations in trying to include China, India, and other major emerging economies in any global climate agreement were virtually identical.

gigawatts (GW) of wind capacity (in terms of name-plate capacity, this is equivalent to 390 coal-fired power stations[27]) and 10 GW of solar, and would shut down 21 GW of coal-fired power stations. Wealth transfers are an essential part of the climate change negotiations, but Podesta conceded that vast wealth transfers from rich countries were not popular in America.[28] This was a problem. Wealth transfers had been an essential part of the original deal that brought developing countries to the table to discuss First World environmentalism.* As another of the conference speakers, Ottmar Edenhofer, told the Swiss *Neue Zürcher Zeitung* in 2010, one has to free oneself from the illusion that international climate policy is environmental policy. Developed countries had expropriated the world's atmosphere; "one must say clearly that we redistribute *de facto* the world's wealth by climate policy."[29]

Edenhofer is an important figure in the climate policy world. A former Jesuit, he is chief economist at Schellnhuber's Potsdam Institute and research director of the Mercator Research Institute on Global Commons and Climate Change, an outfit supported by the Mercator Foundation, which also funded the Essen conference. Since 2008, Edenhofer has been co-chair of the IPCC working group III on climate policy, a position deemed so important that, when first created, the United States insisted it be chaired by a U.S. government official. At Essen, Edenhofer spoke of a strategy of deep decarbonization of the world economy.[30] The term would be picked up and used by the Obama White House when it announced its climate and energy agreement with China in November 2014.[31]

The anticapitalist agenda of the Great Transformation was unmistakeable. Hermann Ott, a former Greenpeace activist, now a Green member of the German federal parliament and also linked with Mercator, talked of the need to "break down the last resistance of the big oil and chemical companies."[32] The German press noted the conference's huge political implications. Essen's local paper, the *Westdeutsche Allgemeine Zeitung*, reported that tackling global warming

* The origins of the deal in the early 1970s struck between First World environmentalism and Third World's development agenda are explored in Chapters 8 and 10 of Rupert Darwall, *The Age of Global Warming* (London, 2013), especially pp. 75–76 and pp. 91–93.

required the reinvention of society.[33] The *Frankfurter Allgemeine Zeitung*, Germany's paper of record, carried a lengthy report that highlighted the conference's call for a reformed culture of citizen participation, new rules and instruments of global governance, and what it called a "cultural filter for perceiving climate change phenomena"—or, in plain English, brainwashing.[34]

There was disquiet in some quarters. Six months later, the essay "Climate Protection Is Killing Democracy" appeared in *Die Welt*. Written by Ulli Kulke, a free-market commentator, it zeroed in on Harald Welzer's "hot question" as to whether authoritarian regimes are better at solving climate change. Welzer's support for democracy is conditional "only when it gets a corrective from the civil society of politically engaged people."[35] Until now, nobody is asking openly for a dictator, Kulke wrote; "only" an authoritarian regime of experts is discussed. In their book, Welzer and Leggewie had expressed very clearly that the green agenda is today more efficiently realized because of centralized political structures. They also argued that to permit continued economic growth was childish. But, asked Kulke, without growth where were all the trillions going to come from to fight climate change? This wouldn't be a problem for authoritarian regimes, Kulke suggested.

Leggewie dismissed fears of an incipient eco-dictatorship. "Climate change means change of culture," he wrote in *Die Welt* in May 2011, requiring a new economic order, new instruments of global governance, and new technologies. "We need a real change in consciousness"—a Marxist throwback to the notion of false consciousness—"and a new stakeholder culture."[36] This cut little ice with the ex-Marxist Gärtner. In a riposte, he noted the East German dictator Walter Ulbricht's infamous assertion: "Nobody has the intention to build a wall" two months before the Berlin Wall went up. "Nobody has the intention to erect dictatorship," Leggewie insists, but then argues for democracy-by-plebiscite, which would be dictatorial for people with dissenting opinions. The eco-transformation was all about the administration of "one world" and the rationing of carbon dioxide and other necessities of life through a world government, Gärtner charged.[37]

The renewed debate had been sparked by prepublicity surrounding the WBGU's 2011 flagship report. *World in Transition: A Social Contract for Sustainability* summarized the themes of the Essen conference. "The extent of the transformation ahead of us can barely be overestimated," the WBGU said. The worldwide remodeling of economy and society were comparable to the two previous great transformations, the Neolithic (the spread of farming and animal husbandry) and the Industrial Revolutions.[38] The previous two were the uncontrolled result of evolutionary change. This one had to be willed by "organizing the unplannable" and achieved within a very tight time frame.[39] The whole economic model of the last 250 years had been based around fossil fuels. "This complex system must now be fundamentally modified."[40] The debate on the limits to growth, begun in the 1970s, had now taken center stage. Urban structures need to be modified, requiring "a determined approach," and habits favoring meat eating also had to be changed.

Without a change in political direction, the WBGU warned that energy consumption could more than double by mid-century. There was a broad, cross-cultural consensus supporting postmaterialistic values. "Politics, economy and society must wholeheartedly embrace long-term orientation."[41] The transformation needs "a powerful state," but the nation-state is not "the sole basis for the contractual relationship." Instead the new contract must take into account other members of the world community and provide "effective, fair global compensation mechanisms."[42] Global governance structures were "the indispensable driving force" of the transformation.[43] The new contract should also include two new protagonists: self-organized civil society (i.e., NGOs) and the community of scientific experts. Then there are the "change agents" across society who could be developed into an effective force. Above all, the new social contract was not so much about what was written down "but rather in people's consciousness."[44] Climate change objectives should be embedded in national constitutions and the interests of future generations safeguarded by having a legislative "future chamber." At the global level, the UN charter should incorporate "planetary guard rails" and keep the rise in global temperature to less than two degrees centigrade.

For America, the message from Essen and the WBGU is stark: to save the planet, you must change your Constitution, subvert your democracy, and accept that the Declaration of Independence is now an anachronism in the transformational struggle against climate change.

3

Northern Lights

> The Swedish Social Democrats interpret the human being exclusively in behavioristic terms. . . . Like the orthodox parties of the Communist world, the Swedish Social Democrats have acted on their belief by manipulating the environment of their citizens in order to create the new man for the new society.
>
> *Roland Huntford*[1]

The two-degree-above-preindustrial-temperatures planetary guardrail would have surprised the first scientist to quantify the greenhouse effect on atmospheric temperature. Svante Arrhenius was born in 1859 in a small village 50 miles from Stockholm. Something of a child mathematical prodigy, Arrhenius grew up to become a physical chemist of repute. In the 1890s, his curiosity was piqued by research carried out by John Tyndall, who speculated that the succession of ice ages and interglacials was caused by changes in the amount of carbon dioxide in the atmosphere.

In 1896, Arrhenius produced the first of two papers. It estimated that a doubling of carbon dioxide would lead to a five to six degrees centigrade (nine to eleven degrees Fahrenheit) temperature increase, but ten years later reduced his estimate to around two degrees. Sweden was a poor country. It had little coal, and people relied on burning wood to keep warm. Far from being alarmed at the prospect of global warming, Arrhenius wanted to increase the use of fossil fuels to help make the Swedish climate milder, which was harsher than today. Arrhenius underestimated the growth of carbon dioxide emissions and thought it would take 3,000 years for temperatures to rise by three to four degrees. He regretted that the beneficial effects of warming would only be enjoyed by our remote descendants.[2]

Arrhenius also believed in technology and progress and was appointed to a government commission to investigate the potential of hydroelectricity. As a pillar of Sweden's scientific establishment, he took a leading role in turning Alfred Nobel's financial legacy into the science prizes that bear Nobel's name. Arrhenius had to wait two years before being awarded the 1903 prize for chemistry.

Belief in technology and progress had a darker side. In 1909, Arrhenius became a board member of the newly formed Swedish Society for Racial Hygiene. Following lobbying by Arrhenius, in 1922, the Swedish government established the Statens institut för rasbiologi (the State Institute for Racial Biology) in Arrhenius's hometown of Uppsala to study ways of improving racial characteristics through selective breeding. According to a 2010 Royal Society paper,

> The Swedish eugenics network may have been relatively small but it was nevertheless historically significant because of its intimate ties with that part of the German eugenics movement that would shape Nazi biopolitics.[3]

As the world's first government-funded eugenics research institute, it attracted a number of German academics, including the notorious racial biologist Hans F. K. Günther, later known as the Nazi Race Pope, who regularly lectured at the race biology institute before becoming the first professor of racial biology at the University of Jena.

The parliamentary bill creating the race institute had been introduced by the psychiatrist and Social Democratic member of parliament (MP) Alfred Petrén, who, like Arrhenius, was a member of the Society for Racial Hygiene and was passed by Sweden's first Social Democratic government. The Swedish government wanted to develop the virtually unpopulated far north of Sweden, home to some 50,000 Lapps or Sami people, nomads who raised reindeer and hunted moose. The Racial Biology Institute's first director, Herman Lundborg, a well-known anti-Semite, was happy to oblige. In an early example of policy-driven research, the Institute set out to find evidence that the Lapps were an inferior people to the Swedes. The Institute also gathered statistics and photographs to assess the racial makeup of 100,000 Swedes, which provided the basis of Lundborg's textbook, *Swedish Racial Studies*, for upper secondary schools.[4]

After being in government on and off in the 1920s, the Social Democrats held power continuously from 1932 to 1976. Modern Sweden is their creation and their creature. In *The New Totalitarians* (1971), Roland Huntford, who was *The Observer*'s Scandinavian correspondent in the 1960s, points out that of all socialist parties, only Sweden's Social Democrats could trace a direct and undefiled descent from Karl Marx himself. Of all countries, Sweden under the Social Democrats had come closest to realizing Aldous Huxley's nightmare of *A Brave New World*, Huntford argued. Swedes were submissive to authority and deferential to experts, which gave Sweden's technopolitical establishment a "singularly malleable population to work with, and has been able to achieve rapid and almost painless results."[5] The system created by the Social Democrats, Huntford wrote,

> has proved to be an incomparable tool for applying technology to society. They have altered the nature of government by making it a matter of economics and technology alone.[6]

Population loomed large in Social Democratic thinking. At the beginning of the twentieth century, the average Swedish family had four children, falling sharply in the next two decades. By 1934, Sweden had the lowest birth rate in the world.[7] The same year, the leading architects of the Swedish welfare state, the husband and wife partnership of Gunnar and Alva Myrdal, wrote down their thoughts on the subject. "Population is the connecting subject," Alva recalled,

> but from that point of view we deal with all social problems: from housing to individual qualities, sterilization, abortion, "the family in the world," nursery schools, equal wages, school feeding, socialization of medicine, Malthus, the eighties, individualism etc. etc. . . . We are going to get enemies in all camps in the social field but I think there will be a revival of discussion along these lines.[8]

In Chapter 7 of *Crisis in the Population Question*, the Myrdals discussed social policy and the quality of the population. If the population is growing, quality is not important, but if the population is stagnant or declining, then there is a need to create a higher-quality population. With technology and automation, the intellectual and moral qualities of workers become more important. There is no place for dull, stupid people in a modern, depatriarchalized system, which also requires women to be integrated into the workforce. They wrote about sterilization policy as the intersection of two lines—the racial hygienic on the one hand and social pedagogic (the ability to change people's behavior, for example by taking youngsters into the care of the state) on the other. The first criterion was "difficult," the Myrdals wrote, as it required determining which individuals should be regarded as carriers of poor genetic material. Deciding on the basis of near-100 percent inheritability of psychological and physical defects would imply the grounds for sterilization would be very restricted, although advances in genetics would improve understanding of the risk of heritability. Theirs was an aggressive form of natalism in which reproduction of low-quality people was to be discouraged by the state and sterilization was to be one of the tools. Clever people would be encouraged to reproduce by giving them generous state handouts, better housing, subsidizing their rents, free medical care, and free school meals. Both parents should work and the patriarchal family structure broken down.

The Myrdals' book had an immense effect on the new Social Democratic government by defining a mission for the state and the party to provide a cradle-to-grave welfare system. Soon after, the state was running the nurseries, and the party's funeral organization was burying and cremating around half Sweden's dead. The Myrdals had a materialist, mechanistic view of people. They were the archetypal social engineers. As distinct from the early eugenicists, they saw population less in biological terms than as a "mathematical or physical quantity."[9]

Despite Alva's expectation about the book creating enemies, the Myrdals reduced their number by deleting sections on eugenics from the English edition, while copies of the Swedish first edition became something of a collector's item. This helped remove a potential obstacle to the Myrdals' stellar postwar reputation, especially in the United States. "Gunnar Myrdal taught me more about economics than anyone else in his generation," J. K. Galbraith wrote in 2005. "He is more relevant now than ever."[10] In 1974, Gunnar was awarded a Nobel Prize in economics. Showing a rare flash of humor, the Nobel committee made him share it with Friedrich von Hayek. Gunnar repaid the compliment in his prize lecture, which barely touched on economics as conventionally understood but showed why he was a darling of the American left. He attacked America for "the illegal, immoral and ruthlessly cruel" war in Vietnam and made a plea for vastly increased foreign aid. America's declining aid program was driven by military and strategic considerations, while Sweden was increasing foreign aid by 25 percent a year. He spoke of a population explosion in developing nations and of the tremendous difficulties in bringing down the birth rate. There should be a new world order, which would start with people in rich countries eating less meat, and he talked of the rationality of a more frugal lifestyle.[11]

Two years later, Gunnar lost his sense of humor when the prize committee awarded the 1976 economics Nobel to Milton Friedman. He said the economic Nobel should be abolished, and he was sorry he had accepted it.[12] Honor of sorts was settled when Alva was awarded the 1982 Nobel Peace Prize for her work in the antinuclear, anti-West peace movement. She showed little hesitation in accepting it, even though Henry Kissinger had been awarded the Peace Prize nine years earlier.

In 1933, the year before the Myrdals put pen to paper, the principle of voluntary consent to sterilization was circumvented by legalizing compulsion in cases where the patient was incapable of giving consent. A government commission on which Gunnar served (Alva was a consulting expert) led to a widening of the grounds for sterilization. Though advocating sterilization of the deficient, he wrote to a friend that the commission's final report had a "smell of Nazism" about it.[13] More generally, according to Huntford, "Nazi thought, often incognito, permeated Swedish life."[14] Commenting on Nazi sterilization policies, the commission wrote, "To admit sterilization without consent to the extent of the German law would probably be inconsistent with the Swedish conception of justice"—hardly an emphatic rejection of Nazi practices.[15] Speaking in 1942, Nils von Hofsten, a zoologist who was Lundborg's deputy at the race institute and an expert to several parliamentary committees,* declared, "sterilization is such an important operation that the individual should not be allowed to decide the matter for himself."[16] It took a communist member of parliament to speak out against such tendencies, which, he said, "bear an unpleasant resemblance to the improvement of the race which one seeks to achieve

* It might appear repellent that on matters of human population the Swedish parliament deferred to the views of a zoologist. In our own time, Paul Ehrlich, an entomologist specializing in the study of butterflies, is also considered an expert on human demography and sustainable development.

in some totalitarian countries by means of the scalpel."[17] Indeed the 1941 law on sterilization further expanded its reach to include those with "an anti-social way of life."[18]

Altogether, 67,000 Swedes were sterilized under these laws. Behind this number lurks a massive gender imbalance. Ninety percent of them were women, indicating that there was more behind the policy than was actually said.[19] A common, though medically risky, procedure was a combined abortion and sterilization, accounting for 60 to 70 percent of sterilizations between 1951 and 1955.[20] A similar gender bias is also evident in Nazi public health policies, which overtly targeted women as bearers of the racial "germ plasm."

There was also—for the most part unarticulated—an anti-Semitic component to Sweden's eugenics policy. Germans did not need visas to enter Sweden, but in 1938, the Swedish foreign ministry informed its German counterpart that Sweden would impose visa restrictions on all Germans unless the German authorities stamped the letter J in the passports of German Jews. Two weeks later, the Swiss government made a similar request. Joachim Israel, an exceptionally fortunate German Jew who managed to settle in Sweden just as the door was closing, wrote in his memoir,

> The Nazis were quite willing to fulfil these demands and surely understood that the Swedish and Swiss requests were confirmation of the correctness of their anti-Semitic policies.[21]

On September 9, 1938, Sweden introduced the *Gränsrekomendationssystemet* (Border Recommendation System), a bureaucratic term designed to sanitize its intent, making it virtually impossible for Jewish citizens of the German Reich to enter Sweden. Israel writes in his memoir that the year he had the J stamped in his passport was the same year "in which the responsible Swedish minister, the brother of the prime minister, issued secret orders that Swedish border guards should turn away all unauthorized Jews who tried to cross the border and send them back to Germany."[22] Solutions to Sweden's population crisis—the one ostensibly identified by the Myrdals four years earlier—were presenting themselves in growing numbers at its borders. The actions of Sweden's Social Democrats revealed that they preferred racial purity over the population growth that the architects of its welfare state said was needed.

4

Europe's First Greens

When man attempts to rebel against the iron logic of Nature, he comes into struggle with the principles to which he himself owes his existence as a man.

Adolf Hitler[1]

The racial aspects of Swedish Social Democrats' brand of modernizing technocracy were implicit. Across the Baltic, in Nazi Germany, race and biology were fundamental components of an ideology that rejected capitalism and industrial society. In his 2006 history of the Third Reich, economic historian Adam Tooze writes of the "disturbing effect" of viewing the period from the perspective of the early twenty-first century,

> rendering the history of Nazism more intelligible, indeed eerily contemporary, and at the same time bringing into even sharper relief its fundamental ideological irrationality.[2]

Similarly, in his book on the Nazi war on cancer, Robert Proctor comments that "the barriers which separate 'us' from 'them' are not as high as some would like to imagine."[3] This is especially so when it comes to the relationship between modern industrialized society and nature, that is to say, the core ideological preoccupation of modern-day environmentalism. "Some of the themes that Nazi ideologists articulated bear an uncomfortably close resemblance to themes familiar to ecologically concerned people today," Janet Biehl and Peter Staudenmaier wrote in 1995 from an anarcho-leftist perspective.[4]

At the root of Nazi philosophy is the idea that humans and nations are compelled to obey the laws of nature. The quotation at the head of this chapter is from a passage in *Mein Kampf* on nation and race (Volume I, Chapter 11). In reproduction, separation of species—and therefore of races—was a biological imperative, Hitler wrote. The point is not whether Hitler got the biology right or wrong (he got it wrong), but the contention that the history of human societies (which in Hitler's biological categorization means race) is determined by biology and nature. The text continues, "Here, of course, we encounter the objection of the modern pacifist, as truly Jewish in its effrontery as it is stupid! 'Man's role is to overcome Nature!'"[5]

The claim that historical development is subservient to natural processes is fundamentally opposed not only to the humanism inspired by the Greeks and to Judeo-Christianity but also to the Marxist left, which saw the laws of history working through the socioeconomic categories of class, not through biological ones. For Marxists, the laws of history are human. Mankind's capacity to overcome nature is what Marx and Friedrich Engels saw as capitalism's greatest achievement. "What is impossible for science?" Engels—who was not a Jew—once exclaimed.[6]

Mankind's subservience to the commands of nature provides the connecting thread between Nazism and modern-day environmentalism and represents a radical rejection of the Enlightenment's belief in progress. It is what separates the New Left and the modern left's softer variants from their predecessors and leaves supporters of capitalism and markets as the last redoubt of belief in the potential of mankind's unfettered material progress.

"This striving towards connectedness with the totality of life, with nature itself, a nature into which we are born, this is the deepest meaning and the essence of National Socialist thought," wrote Ernst Lehmann, a professor of botany, in 1934, who characterized National Socialism as "politically-applied biology."[7] Sixty percent of all biologists joined the Nazis. As Proctor points out, "Nazism took root in the world's most powerful scientific culture, boasting half of the world's Nobel prizes."[8] The culture that gave rise to it existed before the rise of the Nazis. Though dormant in the three decades after 1945, it would reenter German politics in the late 1970s with environmentalism and anticapitalism as two sides of the same coin.

In 1917, Fritz Lenz, a geneticist, wrote the essay "Race as the Principle of Value: On a Renewal of Ethics." Hitler's library contained an offprint of the essay. Lenz argued that the growth of technology had brought about an alienation from nature. When it was republished in 1933, Lenz boasted that the essay "contained all the main features of the National Socialist worldview."[9] Urbanized industrialized man's breach with nature is a recurrent theme. In 1932, the surgeon and cancer specialist Dr. Erwin Liek, widely credited with being the father of Nazi medicine, wrote a book arguing that cancer was a disease of civilization.[10]

Cultural hostility to urban civilization in Germany predates Nazism, and not all those who articulated it became Nazis. In 1913, the philosopher Ludwig Klages, a conservative later criticized by the Nazis, wrote the essay "Man and Earth." Staudenmaier detects in it virtually all the themes of today's environmental movement. Progress was "a sick, destructive joke," Klages wrote, its final goal being

> nothing less than the destruction of life. This destructive urge takes many forms: progress is devastating forests, exterminating animal species, extinguishing native cultures, masking and distorting the pristine landscape with the varnish of industrialism, and debasing the organic life that still survives.[11]

In a remarkable passage anticipating twenty-first-century ecotourism, Klages decried "the hypocritical 'nature feeling' of the tourist trade" and "the devastation which its 'exploitation' of remote coastal regions and mountain valleys leaves in its wake."[12] Such was the modernity of Klages's denunciation of capitalism, consumerism, and economic utilitarianism that in 1980 the essay was republished to accompany the formation of the German Greens as a political party.[13]

The trigger for Hitler's political awakening appears to have come from a 1919 lecture by the economist Gottfried Feder, a founder of the Deutsche Arbeiterpartei (DAP), the German Workers' Party. "As I listened to Gottfried Feder's first lecture about the 'breaking of interest slavery,' I knew at once that this was a theoretical truth which would inevitably be of immense importance for the future of the German people," Hitler wrote in *Mein Kampf*, the solution being "the sharp separation of stock exchange capital from the national economy."[14] According to Hitler, Feder had exposed "with ruthless brutality" the economic character of stock exchange and loan capital—"its eternal and age-old presupposition which is interest."[15]

From this economic critique emerged what Tooze argues is the originality of National Socialism. Hitler's hatred of Jews led him to reject accommodation with a global economic order dominated by affluent English-speaking countries—manifested for him in Wall Street Jewry and the Jewish media—and mount an epic challenge to overthrow it.[16] "Right after listening to Feder's first lecture, the thought ran through my head that I had now found the way to one of the most essential premises for the foundation of a new party," Hitler wrote.[17] In September 1919, Hitler joined the DAP as its fifty-fifth member. The following February, the German Workers' Party added "National Socialist" to its name, becoming the NSDAP—the Nazi Party.

The Nazis' profound hostility to capitalism and their identification with nature-politics led them to advocate green policies half a century before any other political party. As an approximation, subtract Nazi race-hate, militarism, and desire for world conquest, then add global warming, and Nazi ideology ends up looking not dissimilar to today's environmental movement. From the vantage point of the second decade of the twenty-first century, it might come as a shock but should not surprise that Hitler and the Nazis were the first to advocate large-scale renewable energy programs.

1932 was a crucial year for the Nazis. The stock market had crashed in 1929 and the Great Depression fatally undermined the parties supporting the Weimar democracy. The first round of the presidential election was scheduled for March 13. Hitler was the leading candidate to defeat incumbent Field Marshal Paul von Hindenburg. Two and a half weeks before the election, the Nazis' newspaper, the *Völkischer Beobachter*, carried a long article reporting a "sensational speech" on plans for huge 70- to 90-meter-high (230–295 feet) windmills to generate "huge amounts of cheap energy."[18] In the speech, Hermann Honnef of Berlin's Institute of Physics, explained how electricity from the towers could be combined with hydropower, which would be responsible for delivering base-load electricity, while surplus electricity from windfarms located along the coast could be stored by producing "very inexpensive" hydrogen—a green dream that to this day remains just that.

Headlined "National energy policy," the article claimed that wind power could transform the economy and give employment to millions of Germans ("green jobs," in today's political parlance). There would be greater fuel economy through lightweight cars and trains, and tariffs would encourage car drivers to switch to rail. Cars would also use domestically produced synthetic fuels. The program would completely change the whole economy, the article concluded, improving money circulation and reducing imports and cost burdens in one of the first descriptions of "green growth," that alchemical process popular with governments in the twenty-first century whereby making the production of useful energy less efficient would help economies grow rather than do the opposite.

The same year, the party's economic spokesman, Franz Lawaczeck, a founder of the Nazi architects and engineers association (Kampfbund Deutscher Architekten und Ingenieure), wrote a book, *Technik und Wirtschaft im Dritten Reich. Ein Arbeitsbeschaffungsprogramm* (Technology and the Economy in the Third Reich—A Program for Work). An early member of the NSDAP, Lawaczeck was an inventor of hydropower turbines (the U.S. Patent and Trademark Office granted him a patent for a turbine/compressor in 1917[19]). Lawaczeck was on the left wing of the party and associated with Feder, Julius Streicher, Heinrich Himmler, and Joseph Goebbels.

Technology and the Economy is both an anticapitalist tract and a blueprint for a renewable energy future. Lawaczeck extolled the corporatist state that had been destroyed by egoistic liberal capitalism and denounced foreign trade as only benefitting traders.[20] The capitalist system was an experiment in liberal thinking that had led Germany to catastrophe. The experiment was now over and everyone could see the consequences.[21] Like many Nazis, Lawaczeck was impressed by Stalin's Five Year Plans. In the race between liberal and state capitalism, the Bolsheviks would win, Lawaczeck believed.[22]

On energy, Lawaczeck noted that coal had changed the world and attained an overwhelming dominance. Because it was so cheap, it was used wastefully, and long-term priorities were not taken into consideration. Steam turbines used only one-fourth of the energy content of coal, and locomotives used only one-twelfth. Instead of being used for energy, coal would be more valuable making chemicals and other products. By contrast, hydro and wind power were much more efficient, converting up to four-fifths of their energy inputs into useable energy. Those inputs were from nature, "at hand to use, free of cost, but Man has weighed them down to an incredibly high degree with monetary interest."[23]

Lawaczeck argued that the future of wind power was with very large towers like high radio masts, at least 100 meters (328 feet) high, with 100-meter diameter turbines which, unlike modern turbines, were designed to rotate in a horizontal plane.[24] Lawaczeck conceded that hydro and wind power suffered from intermittency. It was impossible to use hydro in winter or in dry, summer months. The situation was even more difficult for wind because of changes in wind speed. The system therefore had to store energy. Like Honnef, Lawaczeck foresaw energy from wind power being converted and stored in the form of hydrogen.[25] The transformation to a hydrogen society would be an important step toward a new industrial revolution, Lawaczeck wrote. Germany could exploit this technology advantage; hydrogen-powered engines would be more powerful than petrol or diesel. It would be cheaper to manufacture and weld steel, giving German exporters a competitive advantage in world markets and enable Germany to export more and pay off its national debt.[26]

Why, asked Lawaczeck, was coal so dominant in capitalist societies and only five percent of Germany's potential hydro power actually used? The answer lay in the upfront investment cost. Coal and oil cost about two hundred Reichsmarks (RM) per kilowatt of generating capacity, while the cost of investing in hydro ranged from 400 to 1,000 RM, Lawaczeck reckoned. However, modern research suggest Lawaczeck was too optimistic about wind. The theoretical maximum of a wind turbine is not 80 percent, as Lawaczeck thought, but 59.3 percent. In practice, the highest attainable power coefficient is around 47 percent.[27] Because air is a thousand times less dense than water, the kinetic energy that can be taken from wind—even at optimal wind speeds of between 25 and 56 mph—is tiny. Thus the energy density of wind is only one-sixteenth that of hydro and five orders of magnitude (nearly one-millionth) less than petrol.*[28]

Wind, like hydro, requires huge amounts of capital. As Lawaczeck and the Nazis grasped—and modern governments would discover in their turn—the market would not provide the capital if left to itself. Echoing Feder's remark about the "breaking of interest slavery," Lawaczeck argued that interest on money is the greatest obstacle to making wind- and hydropower profitable.[29]

Large-scale deployment of Nazi wind power never got off the

* Petrol has around 70 times the energy density of batteries, which is why the performance of electric-powered vehicles is inferior to petrol-driven ones.

ground. If it had, the Allies would have had an easier task in defeating Hilter. Honnef went on to develop plans for a 250-meter (820-ft.) tower with 25 kW generators to be located in a Berlin suburb.[30] Not to be outdone, steelmaker Krupp came up with its own plan for a 540-meter (1,771-ft.) tower—218 feet higher than the Empire State Building, the world's tallest building at the time. The plans were rejected by the management board of BEWAG, the Berlin electrical utility, on cost grounds.[31] Having failed in Berlin, the wind merchants moved on to Munich and formed the Deutsche Windkraft (German Wind Power) lobby group. Honnef had a meeting with Hitler. It didn't go well; Hitler thought Honnef too aggressive in arguing his case.[32]

How politically important were Lawaczeck's green energy plans to the Nazis? On January 30, 1933, Hindenburg appointed Hitler chancellor and immediately sought new parliamentary elections, which were held on March 5. The Nazis did not produce an election manifesto—widespread violence, intimidation, and systematic repression were their main campaign tools, especially after the Reichstag was burned down six days before polling. Nonetheless, there is evidence of the salience of renewable energy in Nazi thinking around this time. Lawaczeck's book was republished in 1933 by the Nationalsozialistische Bibliotek, an affiliate of the party's publishing house, Eher Verlag. Lawaczeck also allegedly featured in one of Hitler's 1932 to 1934 conversations as recounted by Hermann Rauschning.* Hitler had been asked whether new, revolutionary inventions might be expected. Rauschning pointed to Lawaczeck, saying that such inventions were a thing of the past.

> "Engineers are fools," Hitler cut in rudely. "They have an occasional idea that might be useful, but it becomes madness if it is generalised. Let Lawaczeck build his turbines, and not try to invent industrial booms. Don't get mixed up with him. I know his hobby-horse. This is all nonsense, gentlemen."[33]

* Some historians doubt the veracity of Rauschning's account in *Hitler Speaks*, which was written after he'd fled Nazi Germany. Ian Kershaw, for example, boasts that on no single occasion has he cited it, "a work now regarded to have so little authenticity that it is best to disregard it altogether" (*Hitler 1889–1936: Hubris* [New York, 2000], p. xiv). However, it is hard to see what incentives Rauschning would have to fabricate this episode either to earn his publisher's advance or to promote Anglo-French war aims, and Rauschning would have needed considerable powers of invention to have entirely made it up.

The monologue continued. Hitler contended that bringing forth revolutionary inventions was a matter of willpower.

> In one respect Lawaczeck is right: what was once accident must become planned. We must do away with accident. We can! This is the meaning of the "great works" that states undertake today—not speculators and the bank Jews.[34]

Hitler then went on to prejustify Germany's conquest of Europe ("Germany is Europe"), before abruptly concluding: "Lawaczeck, Feder—they're old women! I have no use for their *bourgeois* wisdom!"[35]

Hitler's outburst did not mean he'd cooled on renewable energy. On August 2, 1941, five and a half weeks after launching his attack on Russia, Hitler spoke to his dinner companions. "Because of the fault of capitalism, which considers only private interests, the exploitation of electricity generated by waterpower is in Germany only in its infancy," he told them. "We shall have to use every method of encouraging whatever might ensure us the gain of a single kilowatt. . . . Coal will disappear one day, but there will always be water." Hitler then speculated about possible discoveries of oil and methane before concluding, "the future belongs, surely, to water—to the wind and the tides."[36]

Shortly after Hitler came to power, Lawaczeck lobbied the Reich Ministry of Finance to support wind power. Ministry bureaucrats were skeptical about the costs but provided funding for technical research.[37] In 1935, the wind projects were scaled down because of technical risks. But the finance ministry did give money for a small, 80-meter (262-ft.) tower with a 50-meter turbine, specifying a target cost of 165 RM per kW.[38] Although the rearmament lobby gained the upper hand in pressing for more coal-fired plants, Nazi funding of wind power research continued until 1944 under the auspices of the Reich Windpower Study Group.[39]

Renewable energy was not the only thing that made Hitler and the Nazis Europe's first greens. As is well known, leading Nazis from Hitler down were vegetarians. They argued that meat eating involved inefficient use of land. It took approximately 90,000 calories of grain to produce 9,300 calories of pork. In his Four Year Plan, Hermann Göring said farmers who fattened cattle on grain that could have been used to make bread were "traitors."[40] Himmler blamed commercial interests, using a startlingly modern line of argument. "We are in the hands of food companies, whose economic clout and advertising make it possible for them to prescribe what we can and cannot eat."[41] On his orders, the SS built huge greenhouses at the Dachau concentration camp to grow homeopathic ingredients to be tested on inmates. Around 8,000 prisoners worked in Dachau's industrial-scale greenhouses. In addition to Himmler, leading members of the regime, including its deputy Rudolf Hess and agriculture minister Walter Darré, were fervent supporters of organic farming.[42]

Nazi public health policies incorporated Liek's view that cancer was a disease of civilization—that is, caused by lifestyle choices and industrial pollution. Nazi slogans proclaimed, "Your body belongs to the nation!" and, according to Proctor, Nazi philosophers and politicians admiringly contrasted the National Socialist notion of "health as a duty" with the Marxist one of "the right to do whatever you want with your own body."[43] As Hitler once put it, "Why nationalize industry when you can nationalize the people?"[44] On another occasion, Hitler told an aide that "reforming the human lifestyle" was more important to him than politics.[45]

The Nazi war on cancer thus focused on prevention rather than cure (basic cancer research suffered as Nazi race laws banned Jews from being employed by the state) and the Nazis ran a draconian antismoking campaign. Proctor argues that German epidemiologists discovered the link between tobacco smoking and lung cancer a decade before American and British researchers. Despite being a militant antismoker—Hitler once attributed giving up cigarettes to his being the salvation of the German people[46]—the regime failed to stop the rise in German tobacco consumption (per capita cigarette consumption doubled between 1935 and 1940), although Proctor argues that what he calls Nazi paternalism (in Nazi ideology, women were bearers of the racial germ plasm) diverted the growth in smoking away from women.[47]

Energy independence was also a key Nazi goal, justifying the bailout of one of Germany's leading companies. IG Farben was the first company to place a big bet on oil running out when, in 1926, it invested 330 million RM in the world's first coal hydrogenation facility to transform coal into petroleum at Leuna, near Leipzig.[48] The timing was dreadful. In the late 1920s, a new oil field was found in Oklahoma. Then, in October 1930, wildcatters found the Black Giant in East Texas—the largest oil field discovered until then. Instead of selling for a dollar a gallon, as a U.S. senator had predicted a few years earlier, the price of gasoline fell to as low as nine cents a gallon.[49] Carl Bosch, IG Farben's presiding genius—he won the 1931 Nobel Prize for chemistry—lobbied the Berlin government for protection. By mid-1931, Germany had Europe's highest tariffs on imported oil.[50] Although the Nazis were advocates of energy independence, Hitler also liked fast cars. IG Farben made a 400,000 RM donation to the Nazis' March 1933 election campaign in the largest single campaign contribution to the Nazis' coffers.[51] It paid off. In August, the Reich Ministry of Economic Affairs, where Feder was an undersecretary, wrote to the company, "I have declared to the importers, what guarantees can you give me for the maintenance of world peace?" The ministry committed to promote fuel production from domestic raw materials and provide "the necessary price and sales guarantees."[52]

The deal that kept IG Farben's Leuna facility afloat wasn't a great deal for German motorists. But there weren't many of them on the road because motoring was so expensive. In 1934, there was 1 car for every 5 Americans; 1 for every 22 in France and 27 in Britain. For Germany, the figure was 1 car for every 75 Germans.[53] According to Tooze, taxes and the legal mandate to add domestically produced ethanol (the Nazis were first with those too) doubled the price of petrol. By the late 1930s, the high price of petrol meant that a 100-mile journey cost an entire day's wages.[54] Keeping Leuna in business and making motoring more affordable required cutting the cost of cars. Speaking at the 1936 International Motor Show, Hitler attacked the car industry. It was unacceptable that Germany, with half the population of the United States, should have one-fiftieth as many motor vehicles.[55]

General Motors' Opel subsidiary produced a new range of small family cars, but the 1,450 RM price tag was too high for Hitler, who wanted to drive the price down to 1,000 RM. The private sector couldn't build it, which is where Ferdinand Porsche came in. Together with soft financing from the German Labor Front, Porsche would design and build a people's car, the Volkswagen. Over a quarter of a million people deposited 275 million RM to get on the Volkswagen waiting list, but none of them got a car. The entire VW output was delivered to the military.[56] Building autobahns was not for civilians driving VW Beetles but to enable troops to be moved from the eastern to the western border of the Reich in two nights' hard driving.[57] Even here, environmental considerations had a place. Fritz Todt, in charge of the massive project, declared: "The German highway must be an expression of its surrounding landscape and an expression of the German essence."[58] The Nazis passed hedgerow and copse protection ordinances and required new tree plantations to include deciduous trees. At the height of the war, Hitler acted to protect wetlands.[59]

In the dying days of the regime, Hitler chalked up one final environmental accomplishment. By the spring of 1945, the Rhine was running clear for the first time in generations. "There were no factories left in operation to pollute it."[60] It was the realization of a green dream, of sorts.

After the end of Hitler's war, the Allies ignored German research on tobacco smoking and lung cancer. There was, however, a channel through which Nazi environmental protection ideas found their way across the Atlantic. In 1923, Wilhelm Hueper had left his native Germany and settled in America, eventually becoming a pathologist at the University of Pennsylvania's Cancer Research Laboratory. Nine months after Hitler came to power, Hueper wrote to the new government asking for a job and expressing a desire to renew his bonds with his native culture and people.[61] Hueper returned to the United States in 1934 and in 1938 became the first director of the Environmental Cancer Section of the National Cancer Institute. But Hueper's greatest impact on environmental health policy was through the foundational text of postwar environmentalism, making eight appearances in the cancer chapter of Rachel Carson's *Silent Spring*. Under the menacing chapter heading "One in Every Four" (in 1955, the American Cancer Society had predicted that cancer would strike one in four Americans), Carson told a tale of industrial society releasing carcinogens into the environment.

> The most determined effort should be made to eliminate those carcinogens that now contaminate our food, our water supplies, and our atmosphere, because these provide the most dangerous type of contact—minute exposures, repeated over and over throughout the years.[62]

She cited Hueper to argue that reliance wholly or even largely on therapeutic measures "will fail because it leaves untouched the great reservoirs of carcinogenic agents which would continue to claim new victims faster than the as yet elusive 'cure' could allay the disease."[63] Indeed, some of the language of *Silent Spring* could have been written three decades earlier in Germany—references to chemicals shattering the "germ cells" and altering "the very material of heredity" and to "the human germ plasm."[64] In a Liek-like sentence, Carson wrote of "the tide of chemicals born of the Industrial Age" bringing about drastic change in the most serious public health problems.[65]

Were Liek, Hueper, and Carson right? Omitted from Carson's reservoir of carcinogens is the one that has killed far and away the largest number of people—the humble tobacco leaf. A 1981 paper for Hueper's institute written by epidemiologists Richard Doll and Richard Peto concluded:

> Examination of the trends in American mortality from cancer over the last decade provides no reason to suppose that any major new hazards were introduced in the preceding decades, other than the well-recognized hazard of cigarette smoking.[66]

Environmental exposures were substantially less important, the paper found. Carson's war on industrial society was, like the Nazis' before it, motivated by ideological zealotry, not by evidence.

Present-day resonances with National Socialism have divided historians. Proctor criticizes those who find a fascist danger inherent in any state-sponsored public or environmental health protection policy.[67] The Nazi war on cancer demonstrates that "good" science can be pursued in the name of antidemocratic "ideals" and Proctor suggests that we should understand the fertility of fascism and not just its cruelty. "Public health initiatives were pursued not just *in spite* of fascism, but also *in consequence* of fascism" (Proctor's emphasis).[68]

Often it is those on the left who understand the importance of ideology. Staudenmaier, for one, will have nothing to do with any apologetics for National Socialism.

> To make this dismaying and discomfiting analysis more palatable, it is tempting to draw precisely the wrong conclusion—

> namely that even the most reprehensible political undertakings sometimes produce laudable results. But the real lesson is just the opposite: Even the most laudable of causes can be perverted and instrumentalized in the service of criminal savagery.[69]

In his 1999 essay "Antimodernism, Reactionary Modernism and National Socialism," Thomas Rohrkrämer provides further unsettling analysis. He highlights historians' widespread refusal to accept that National Socialism existed within the framework of modern societies and displayed specifically modern features.

> Instead of distancing modernity from National Socialism, we should learn to accept that it was by no means a necessary, but was a possible development within modernity. In that sense, National Socialism shows modernity's most fatal potential.[70]

Nazism arose out of a specific set of historical circumstance, and its trajectory would have been entirely different without the man who put the National Socialist into the German Workers' Party. Yet Rohrkrämer's warning should not let the uniqueness of those circumstances blind us to what he calls "modernity's most fatal potential." Coming after the Nazi catastrophe imposes on us an obligation to understand not only National Socialism but also, to borrow from Staudenmaier, its ideological continuities and common genealogy in whatever form. That requires us to examine critically those things the Nazis and modern environmentalists have in common—what they were for and what they are against. And it means using as our yardstick what they would sacrifice in order to achieve the goals they share. Of these, the most precious is the liberty of the individual.

5 Intellectuals, Activists, and Experts

> Most civilizations have disappeared before they had time to fill to the full the measure of their promise.
>
> *Joseph Schumpeter*[1]

> Universities are always oases.
>
> *Hebert Marcuse*[2]

> Make the enemy live up to their own book of rules. You can kill them with this, for they can no more obey their own rules than the Christian Church can live up to Christianity.
>
> *Saul Alinsky*[3]

Hitler's anti-Semitism had one unintended effect: It put the atom bomb beyond his reach. German physics had been contaminated by incomprehensible new ideas such as relativity and quantum theory. These were the result, physicist and Nobel Prize–winner Philipp Lenard said, of a "massive infiltration of the Jews into universities. . . . The most obvious example of this damaging influence was provided by Herr Einstein."[4] The Nazi civil service law purged Jews from employment in public administration, including universities. Warned that Germany was throwing away its scientific supremacy, Hitler retorted: "If the dismissal of [Jews] means the end of German science, then we will do without science for a few years."[5]

Another group of Jewish academics fled Germany for Switzerland not because they were Jews but because of their politics. Within six weeks of taking power, the Nazis closed the Marxist Institut für Sozialforschung (Institute for Social Research), known to the world now as the Frankfurt School. The Institute would have an enormous impact in radicalizing American universities and developing a post-Marxist justification for environmentalism. In so doing, it provided a bridge that would enable environmentalism to make its traversal from National Socialism, where it would constitute a major part of the New Left agenda.

Six months before the members of the Frankfurt School left Germany, the foremost economist of the German-speaking world arrived at Harvard. Joseph Schumpeter had a profound understanding of the propulsive, dynamic nature of capitalism. J. K. Galbraith called him the most sophisticated conservative of the twentieth century.[6] But Schumpeter was darkly pessimistic about its future and foresaw capitalism turning into socialism. While the Frankfurt School furnished intellectuals and activists with weapons in the war against capitalism, Schumpeter described a process ineluctably turning capitalism into socialism. What Schumpeter could not have foreseen—he died in 1950—was the role that environmentalism would play. Beginning in the 1960s, business leaders and opinion formers embraced environmentalist tenets. Environmentalism became the fuel for the engine of capitalism's self-destruction.

It was in America that Schumpeter wrote *Capitalism, Socialism and Democracy*. Best known for his description of capitalism as a "perennial gale of creative destruction," this was a prelude to the book's most audacious claim: Capitalism would decay into socialism not, as Marx believed, because it would fail economically but through its very success.[7]

Schumpeter's analysis of capitalism's demise has two blades. The first is a class-based analysis of the atrophying of the entrepreneurial function. The bourgeoisie depends on the entrepreneur and, as a class, "lives and will die with him."[8] The other is an analysis of the social forces generating ideas hostile to capitalism. The rationalism that produced capitalism goes on to break down the values and institutions on which capitalism depends:

> Capitalism creates a critical frame of mind which, after having destroyed the moral authority of so many institutions, in the end turns against its own; the bourgeois finds to his amazement that the rationalist attitude does not stop at the credentials of kings

> and popes but goes on to attack private property and the whole scheme of bourgeois values.[9]

The later stages of capitalist civilization see the expansion of the educational apparatus, particularly that of higher education. Intellectuals have the function of the anti-entrepreneur. "Unlike any other type of society, capitalism inevitably and by virtue of the very logic of its civilization creates, educates and subsidizes a vested interest in social unrest," he wrote.[10] Yet the capitalist order was incapable and unwilling to "control" its intellectual sector.

> Freedom of public discussion involving freedom to nibble at the foundations of capitalist society is inevitable in the long run. On the other hand, the intellectual group cannot help nibbling, because it lives on criticism and its whole position depends on criticism that stings.[11]

Schumpeter's ten-word aphorism—"Capitalism pays the people that strive to bring it down"—perfectly describes the Frankfurt School.[12] Established in 1923, its founder, Felix Weil, was the son of a grain merchant who used his father's wealth and his mother's inherited fortune to support radical ventures. The original idea of calling it the Institut für Marxismus was abandoned as "too provocative." Nor did Weil want the institute named after him, as he "wanted the Institut to become known, and perhaps famous, due to its contributions to Marxism as a scientific discipline, not due to the founder's money."[13] So Institute for Social Research it became. Thanks to a wealthy benefactor, in 1930 Max Horkheimer—a philosophy professor at Frankfurt University—was appointed the Institute's second director, a position he held until 1953.

Marxist intellectuals of the prewar era needed to explain why the working class had failed to act in accordance with their class interest. The Frankfurt School argued that the proletariat had been diverted from its revolutionary role by mass culture and developed a body of ideas under the banner of "critical theory." Its principal innovation was integrating Freudian psychoanalysis into Marxism: Capitalism embodied Thanatos, the life- and nature-destroying primary drive, and suppressed Eros, the instinct for life. This synthesis would provide the Institute its attack on modern industrial civilization.

The aim of critical theory was not genuine understanding of social phenomena, derided as the "fetishism of facts," but bringing about social change.[14] Truth was not immutable. Each period had its own truth. What is true is whatever fosters social change in the direction of a rational society, Horkheimer wrote in 1935.[15] Disinterested scientific research was impossible in a society in which men were themselves not autonomous and empiricism a capitulation before the authority of the status quo. Critical theory does not set out what a rational, socialist society actually might be and what critical theory stood for. "There is no free society without silence," Marcuse mused.[16] As a clue, that doesn't take us far. In a lecture at the end of his life, the best Marcuse could proffer was an oxymoron and the prospect of a "concrete utopia"—"'utopia' because such a society is a real historical possibility."[17] As Martin Jay, historian of the Frankfurt School, acknowledges,

> dialectics was superb at attacking other systems' pretensions to truth, but when it came to articulating the ground of its own assumptions and values, it fared less well.[18]

Although the members of the Frankfurt School were to lessen their enthusiasm for the Soviet Union, their criticism was muted. "One ought not to ignore the difficulties involved in a Marxist analysis, however heterodox, of communism's failures," in Jay's circumlocution.[19]

Having secured an affiliation with Columbia University, the Institute moved from Switzerland to New York, where it developed a parasitical relationship with its new host country. Not defining what they were for, the Frankfurt School was vehement in what it was against. Critical theory was about harnessing the power of "negative thinking" against America, that "repressive monolith," Marcuse said.[20] Capitalism was blamed for every ill, even communist ones. "I believe that many of the reprehensible things that happen in Communist countries are the result of competitive co-existence with capitalism," Marcuse opined in 1968.[21]

The Frankfurt School's breakthrough in the United States came in 1943, when the American Jewish Council gave it a large grant to study American anti-Semitism. This was not without some irony. The researchers of the Institute for Social Research had been blind to the rise of anti-Semitism in the country they had fled. "The German people are the least anti-Semitic of all," one of them had written ten years after the Nazis came to power. Another confessed in 1970 that "all of us, up to the last years before Hitler, had no feeling of insecurity originating from our ethnic descent."[22]

When *Studies in Prejudice* appeared in 1950, one reviewer acclaimed it as "perhaps an epoch-making event in social science."[23] Of its five volumes, the most famous was *The Authoritarian Personality,* incorporating Theodor Adorno's F-scale, which purportedly measured and categorized personality types according to their susceptibility to authoritarian ideas (the F stands for fascist). Other reviewers were less adulatory. Writing in the *American Journal of Sociology* in 1952, Tamotsu Shibutani identified the Frankfurt School's characteristic mode of argumentation as foreclosing the possibility of debate and the legitimacy of disagreement. "While it may be comforting to regard those whom we dislike as pathological and lend 'scientific' sanction to our condemnation, such a procedure is not always conducive to an impartial analysis and genuine understanding of the phenomenon in question," Shibutani wrote.[24]

In a withering 1957 review, the Chicago sociologist Edward Shils wrote that the Frankfurt School's interpretation rested on images of modern man, of modern society, and of man in past ages that had little factual basis. "It is a product of disappointed political prejudices, vague aspirations for an unrealizable ideal, resentment against American society, and at bottom, romanticism dressed up in the language of sociology, psychoanalysis and existentialism."[25] As Shils explained, the school's decisive step was breaking with the materialism of classical Marxism—"vulgar Marxists," according to Frankfurt School taxonomy—and its urge to dominate the world and nature. Marx had wanted to turn the whole world into a workhouse, Marcuse said later.[26] This opened the way for the Frankfurt School's repudiation of industrialization and adoption of what Shils called an "aesthetic repugnance for industrial society."[27]

> Their earlier economic criticism of capitalistic society has been transformed into a moral and cultural criticism of the large scale industrial society. They no longer criticize the ruling classes for utilizing the laws of property and religion to exploit the proletariat; instead they criticize the "merchants of kitsch" who are enmeshed in the machine of industrial civilization and who ex-

ploit not the labour but the emotional needs of the masses—these emotional needs themselves produced by industrial society.[28]

The Frankfurt School thus furnished the New Left with an indictment of industrial society which explained both why American workers had rejected socialist revolution *and* capitalism's supposed destruction of the environment. Contemptuous of the cultural values of the American working class, the New Left castigated them as accomplices in the emergence of fascism in late-stage capitalist society. This theme was popularized by Marcuse in his 1964 *One-Dimensional Man,* which became a canonical text of the New Left and had considerable impact in West Germany. According to the historian of the German New Left, Paul Hockenos, West German students, like their American counterparts, found a fitting description of their "apolitical, mass-media-fed culture."[29] Dog-eared copies of *One-Dimensional Man* were passed around German campuses. *The Authoritarian Personality* was also "extremely influential."[30]

The notion that America was on a conveyor belt to fascism showed just how little these *émigrés* from Nazi Germany understood about Nazism and capitalism. As Adam Tooze shows, Nazism was conceived as a radical rejection of capitalism and the Anglo-American liberal economic order, which in Hitler's view required Germany to wage a world war to overthrow it. Thus the historical premise of the Frankfurt School was completely false. But what it succeeded in doing was to create an ideological justification for environmentalism to become a central part of the New Left's and the 1960s counterculture's rejection of capitalism—a post-Marxist philippic against consumerism and materialism which until then had principally been the province of reactionary, anticapitalist movements on the right and Far Right, and of National Socialism.

This development was not an accident. In his 2004 book on postmodernism, the philosopher Stephen Hicks argues that by the 1950s, the failure of Marxism to develop according to the logic of traditional theory had reached a crisis. This led to the Far Left in effect agreeing with what the anticapitalist, collectivist right had long argued: "Human beings are not fundamentally rational—that in politics it is the irrational passions that must be appealed to and utilized."[31] At a deep level, the philosophers of the New Left share the nihilism and irrationalism of National Socialism. Both rejected the Enlightenment and belong to the tradition of the German counter-Enlightenment.

Frankfurt School dialectics armed progressives with the linguistic weaponry to fight the culture wars. They perfected the technique of taking two words with antithetical meanings and ramming them together to drain them of positive attributes. "Totalitarian democracy" is one. "Repressive tolerance"—the title of a polemic Marcuse wrote in 1965—is another. The first step in Marcuse's torture of tolerance is to define the circumstance, that is, a socialist utopia, where tolerance is justified.

> Tolerance is an end in itself only when it is truly universal. . . . As long as these conditions do not prevail, the conditions of tolerance are 'loaded.'"[32]

If it protects and preserves a repressive society and thereby "serves to neutralize opposition and to render men immune against other and better forms of life, then tolerance has been perverted."[33] A prerequisite for the strengthening of progressive ideas therefore requires suppression of regressive ones—what Marcuse calls "liberating tolerance," also called repression.[34] "Realization of the objective of tolerance would call for intolerance toward prevailing policies, attitudes, opinions," Marcuse wrote.[35]

Which views should be promoted and which suppressed? As Marcuse put it, "The decisive problem is to determine whether the limitations imposed on the individual are imposed in order to further the domination and indoctrination of the masses, or, on the contrary, in the interest of human progress."[36] Who decides what is progressive? Progressives, of course. The anathematization of their opponents in the culture wars and the delegitimization of disagreement ("climate denier") in the climate wars is a contemporary example of "liberating tolerance."

The university professoriat was the principal channel through which the ideas of the Frankfurt School were spread in the United States:

> Although they wrote and lectured about an intellectual tradition critical of most aspects of US society, scholars of the Frankfurt School were invited into the establishment, earning chairs

at such prestigious universities as Harvard, Yale, Princeton, Cornell, Columbia, Duke, the University of California at Berkeley, and University of Chicago.[37]

Its impact on the radicalization of German students in the 1960s was more direct, especially on the circle of Far Left radicals from which the leadership of the future Greens would emerge. The Frankfurt School's efforts in the United States had been financed by the fruits of American capital. The Rockefeller Foundation paid for Marcuse's study on Marxism in the Soviet Union: "The two societies (both East and West) were actually moving toward each other—bureaucratic, totalitarian civilizations."[38] With the encouragement of postwar American occupation officials and John J. McCloy, the high commissioner for Germany, Rockefeller Foundation dollars helped finance the school's return to Frankfurt. Those radicals included Rudi Dutschke, a student leader and antiwar activist who, in 1968, survived several bullets to his head and body, one of the incidents that year that led to the emergence of a new generation of student radicals—the 1968ers. In a 1968 tract, "On Anti-authoritarianism," Dutschke argued that "a revolutionary dialectic of the correct transitions must regard the 'long march through the institutions' as a practical and critical action in all social spheres."[39] Dutschke started but did not finish that three-decade march, which would culminate in Berlin's first Red–Green coalition. Shortly before his death in 1979 at the age of 39, Dutschke was involved in the preparations to launch the Greens as a new radical electoral force in German politics.

In 1968, as student riots convulsed West Berlin, America, and France, three journalists from the French weekly *L'Express* caught up with Marcuse on the French Riviera. What he told them reveals more about the attitudes of the Frankfurt School and its long-term impact than its formal writings.

L'Express: Can we say for the students who have chosen a doctrine for their revolt that you are their theorist?

Herbert Marcuse: "I have tried to show that contemporary society is a repressive society in all its aspects, that even the comfort, the prosperity, the alleged political and moral freedom are utilized for oppressive ends. . . . It is not merely a question of changing the institutions but rather, and this is more important, of

> totally changing human beings in their attitudes, their instincts, their goals, and their values."

Do you mean that this is fundamentally a humanist movement?

> "They [the students] object to that term because, according to them, humanism is a bourgeois, personal value. It is a philosophy inseparable from destructive reality."

France is very far from that "affluent" society whose destruction you propose which only exists, for better or worse, in the United States.

> "I have been accused of concentrating my critique on the US, and this is quite true. . . . But this is not only because I know this country better than any other; it is because I believe or I am afraid that American society may become the model for the other capitalist countries, and maybe even for the Socialist countries."

Wouldn't a revolution result in exchanging one series of restraints for another?

> "Of course. But there are progressive restraints and reactionary restraints. . . . For example, industries would not be permitted to pollute the air."

Can't this formula be turned against you?

> "It always is. And my answer is always the same. I do not believe that the Communism conceived by the great Marxist theorists is, by its very nature, aggressive and destructive; quite the contrary."[40]

Capitalism, of course, did destroy man and nature. In a lecture in 1979, shortly before his death, Marcuse argued that in advanced industrial society, material satisfaction

> is always tied to destruction. The domination of nature is tied to the violation of nature. The search for new sources of energy is tied to the poisoning of the life environment.[41]

American biologist Garrett Hardin attacked capitalism from a different angle. In 1968, Hardin gave a lecture to the American Association for the Advancement of Science that became one of ecology's canonical texts (*Science* lists over 26,000 citations). "The Tragedy of the Commons" takes a (hypothetical) example of open pasture leading to overgrazing. On the basis of this, Hardin asserted that freedom in a commons brings ruin to all. Adam Smith's invisible hand had to be replaced by government coercion. It was theorizing without facts. As the Nobel economist Elinor Ostrom wrote in a 1999 critique, such a prescription was not supported by extensive research. In fact, there is a wide diversity of institutional arrangements for coping with common-pool resources—when they had not been prevented by central authorities.[42]

Population control was at the core of Hardin's concerns: "Freedom to breed will bring ruin to us all."[43] In an earlier biology textbook, Hardin had asked students to confront the benefits of eugenics in reducing the "supply of mutant types at equilibrium."[44] Should society resort to eugenics? "Having upset the primeval balance of nature by producing Pasteur and all his name symbolizes, mankind must now invent corrective feedbacks to restore the equanimity of life."[45] This involved more than biology; "eugenic measures (of which there is a wide variety) involve some sort of interference with personal freedom."[46] Hardin did not have a high regard for freedom. In his 1993 book, *Living within Limits*, freedom was relabeled "radical individualism," something that was not very old (only dating back three centuries to John Locke, Hardin said) and confined to a minority of the world's population.[47] Hardin quoted Hegel's formulation: "Freedom is the recognition of necessity"—a slogan also used by Engels and drilled into the party faithful across the communist bloc as a core tenet of Marxism-Leninism.[48]

Where to place Hardin on an anti-capitalist left–right spectrum? He was anti-cities ("urbanization may, in the end, prove to be a fatal disease"[49]), against immigration ("rich nations must refuse immigration to people who are poor because their governments are unable or unwilling to stop population growth"[50]) and even justified sun worship ("religions that revolve around sun worship make a sort of sense"[51])—a feature of both the harmless Kibbo Kift movement in England of the interwar years but also of early Nazi race ideology.[52] Wherever his political sympathies lay, they are unlikely to have been with the New Left. But Hardin left a few clues. "Some day, political conservatism will once again be defined as contented living within limits," he wrote in 1993.[53]

Thus environmentalism united enemies of capitalism and the Western ideal of freedom from otherwise mutually antagonistic ideologies. Anarchists were other foes of capitalism who founded environmentalism. The aim of the anarchist movement, Murray Bookchin wrote in an extraordinarily prophetic 1964 essay, was a stateless, decentralized society based on communal ownership of the means of production. Burning fossil fuels showed modern man's disruptive role. The amount of carbon dioxide had increased by thirteen percent since the Industrial Revolution. The growing blanket of carbon dioxide would lead to rising temperatures, more destructive storm patterns, rising sea levels, and, in two to three centuries, the melting of the polar ice caps, Bookchin wrote. The solution was to apply ecological principles to energy generation. "We could try and re-establish earlier regional energy patterns, using a combined system of energy provided by wind, water, and solar power."[54] Wind power was ideal in mountainous areas, and solar devices and heat pumps could provide as much as three -quarters of the energy required to heat a small family house. Bookchin liked tidal power, highlighting a tidal project being built in Brittany. "In time the Rance River project will meet most of the electrical needs of northern France," Bookchin claimed.*

In terms of directly radicalizing American students of the 1960s, neither the Frankfurt School, Hardin, nor American anarchists lit the fuse. If it was anyone, it was the sociologist C. Wright Mills. Texan by birth, Mills combined American practicality with deep immersion in German philosophy. From his time at Columbia, Mills was acquainted with the work of the Frankfurt School and had reviewed one of Horkheimer's books. A bigger influence on Mills was the German sociologist Hans Gerth, who had taken courses at Horkheimer's institute in Frankfurt as well as studying at a rival school of Marxist sociology. "I have been studying, for several years now, the cultural apparatus, the intellectuals—as a possible, immediate, radical agency of change," Mills wrote in 1960.[55] In the Soviet bloc, it was the students and young professionals, the young intelligentsia who were exhibiting signs of breaking out of apathy. "That's why *we've* got to study these new generations of intellectuals around the world as real live agencies of historic change.[56]

Mills combined his sociological insight and his aim of radicalizing

* Bookchin might well have been the first person since the Nazis, but certainly not the last, to have been exceedingly optimistic about the potential of renewables. With an annual output of 540 GWh, Rance generates enough electricity to power fewer than 100,000 homes (based on 2013 average French household consumption of 5,830 kWh per year [Source: World Energy Council]).

American campuses with a book that did just that. *The Power Elite* (1956) describes an America where, underneath the trappings of democracy, networks of men at the top of powerful hierarchies make all the decision that matter: They rule the big corporations, run the machinery of state, direct the military establishment, and occupy the strategic command posts of the social structure.[57] This analysis fired up a group of students at the University of Michigan, where Tom Hayden, who was writing his master's thesis on Mills, formed Students for a Democratic Society (SDS). Hayden drafted the society's manifesto. The Port Huron Statement argued that American corporations were not democratically accountable: "It is not possible to believe that true democracy can exist where a minority utterly controls enormous wealth and power." A new reordering was necessary: "We must consider changes in the rules of society by challenging the unchallenged politics of American corporations."[58]

Hayden pointed progressives to capitalism's weakest point. In *Capitalism, Socialism and Democracy*, Schumpeter argued that dematerialized, defunctionalized absentee ownership does not summon the moral allegiance that private property once did. Corporations—the bourgeois fortress—become politically defenseless. "Defenseless fortresses invite aggression especially if there is rich booty in them. Aggressors will work themselves up into a state of rationalizing hostility—aggressors always do."[59] As the political scientist Jarol Manheim writes in *Biz-War and the Out-of-Power Elite*, for progressives, corporations are the perfect enemy.[60]

Mills died in March 1962, and the SDS produced the Port Huron in June of that year—three months before publication of Rachel Carson's *Silent Spring*. Environmentalism had not featured in Mills's work or in the Port Huron Statement, but thanks to the outpouring of environmental consciousness triggered by *Silent Spring*, the insurgents didn't have to storm the ramparts. Not only did environmentalism unlock the fortress gates; they didn't have to loot and plunder, as Schumpeter had anticipated. The occupants of the fortress threw them a welcome party and wealthy philanthropic foundations—Rockefeller, Ford, McArthur, and Pew—gave them the booty. Corporate and foundation wealth and power were handed to them on a plate.

Already in 1962, the Aspen Institute for Humanistic Studies had run a program on climate in the eleventh and sixteenth centuries—"the subject entailed an enquiry into the ways sharp climatic changes could affect the conditions of life on earth."[61] The Aspen Institute had been founded in 1949 as a celebration of humanism and German culture (1949 was the bicentenary of Goethe's birth). After the death of founder Walter Paepcke in 1960, Aspen's new chairman, Robert Anderson, wanted a new focus. Aspen's climate program fed into a growing preoccupation with environmental issues, what became known as the "outer limits"—the interplay between climate, food supply, and population growth. Anderson, who was also chairman of Atlantic Richfield, helped finance the first Earth Day in 1970 and provided the seed money that got Friends of the Earth started.[62]

In 1969, Anderson hired Joseph Slater to run Aspen. Slater came from the Ford Foundation, where John McCloy had been chairman. Slater was a formidable organizer. With financial backing from the Ford Foundation, he helped start the London-based International Institute for Strategic Studies and wanted Aspen to spawn a similar institution focused on the environment and solving what was viewed as an emerging planetary crisis. Slater headhunted Thomas Wilson from the State Department, where he had failed to put environmental concerns at the forefront of the international agenda. In *The Environment: Too Small a View* (1970), which he wrote at Aspen, Wilson argued that society needed to be "managed."

> The "crisis" has more to do with economic-political-social change than with more and better sewage treatment and smoke abatement, essential though these may be. It is a crisis not just for the environment but for traditions and institutions as well.[63]

A two-day workshop of 75 distinguished experts was held at Aspen in September 1970. It came to the "melancholy conclusion" that modern technology, greedy men, and complacent or inefficient government were threatening the future of a decent and civilized world. "All insist that the human family is approaching an historic crisis which will require fundamental revisions in the organization of society," the *New York Times*'s James Reston reported.[64] None of them had stumbled on any definitive answers, but as Reston noted, the meeting had been grappling with questions left unanswered for over 2,000 years. For the experts at Aspen, the reality of a gathering environmental storm was becoming an article of faith.

Slater was friends with Sweden's ambassador to the UN, who in May 1968 had formally proposed to the UN General Assembly that the UN should convene, for the first time, a major conference on the environment. The proposal went forward. It would be held in Stockholm in June 1972. The Aspen Institute's activities were then geared around the impending UN conference. Wilson established the Aspen Environment and Quality of Life Program. Leading environmentalists who appeared at Aspen include Lester Brown (the Rockefeller Foundation financed Brown's Worldwatch Institute and subsequently the Earth Policy Institute) and the two principal architects of the Stockholm conference—Barbara Ward and Maurice Strong.* Wilson acted as Strong's personal consultant. He wined and dined corporate leaders in Paris and New York, ending up with the International Chamber of Commerce formally pledging support for the UN program on the environment. At the conference itself, the Aspen Institute hosted a Distinguished Lecture Series financed by the International Population Institute. Speakers included Gunnar Myrdal, Barbara Ward, and the Club of Rome's Aurelio Peccei, who had commissioned *The Limits to Growth*, which was published two months before the conference. "Likely to be one of the most important documents of our age," Anthony Lewis wrote in the *New York Times*, showing "the complete irrelevance of most of today's political concerns."[65]

1972 was the year postwar environmentalism came of age. In the 1960s, environmentalism had united otherwise antagonistic anticapitalist ideologies because capitalism was allegedly destroying nature. This accomplishment would have counted less had captains of industry and finance not also subscribed to the belief that unfettered capitalism threatened the future of the planet, fears that were amplified by a cast of opinion leaders and a media uncritically accepting dire forecasts of impending doom. From then on, the environmental movement had the institutional support of the United Nations and its agencies. That year's Stockholm conference on the environment was the first in a series stretching from the 1992 Rio Earth Summit to the 1997 Kyoto climate conference, from Copenhagen (2009) to Paris (2015) and beyond.

Belief in planetary eco-doom required specific crises to focus po-

* See Chapter 8, Rupert Darwall *The Age of Global Warming: A History* (London, 2013).

litical attention and generate the requisite level of alarm. The first of these was the acid rain scare. Acid rain was the dress rehearsal for global warming. The difference is we know how it ended. Traversing the series, committed eco-travelers always looked forward to the next eco-catastrophe, never back to see if the previous one had actually materialized. If they had, they would have seen not the dying trees and dead lakes of their imagining but thriving forests and woodlands and healthy rivers and streams.

6

Raindrops

> It was the prediction that motivated people to check for damage; research was intended in part to test the prediction, and in part to stimulate action before it was too late to stop. . . . Still, if the point were to prevent damage before it happened then such arguments were necessarily speculative. A careful scientist would be in a bit of a bind: wanting to prevent damage, but not being able to prove that damage was coming.
>
> *Naomi Oreskes and Erik Conway on acid rain, 2010*[1]

Acid rain was the reason Sweden wanted the UN conference on the environment. In 1967, Svante Odén, a Swedish soil scientist at the Agricultural College of Uppsala, had produced a complete theory of acid rain. Odén wrote a sensationalist article in the leading Swedish daily *Dagens Nyheter* on a "chemical war" between the nations of Europe.[2] Outside Scandinavia, acid rain was not considered important. At the suggestion of Aspen's Thomas Wilson, Barbara Ward and micro-biologist René Dubos wrote *Only One Earth* as a curtain raiser to the conference. Climate change—warming and cooling—got five and a half pages.[3] Acid rain wasn't mentioned, so Sweden took matters into its own hands. At the end of November 1970, the UN secretary-general asked member governments to present environmental case studies. Within a week, the Swedish government replied with a proposal on acid rain and convened a group of experts chaired by the head of Stockholm's International Meteorological Institute, Bert Bolin.

Acid rain was not only a precursor of global warming; it was the prototype. Both mobilized the same constituencies—alarmist scientists, NGOs, and credulous politicians—amplified by sensationalist media reporting. The target was the same—fossil fuels, especially coal. For both, scientists could point to a scientific pedigree stretching back into the nineteenth century—"acid rain" was first coined by the British scientist Robert Angus Smith in the 1870s. Both originated in Scandinavia and were taken up by Canada. The same scientists were active in both. The first chair of the IPCC, Bert Bolin, wrote the first governmental report on acid rain. Both had the capacity to induce hysteria—if anything, more extreme in Germany at the height of the acid rain scare in the early 1980s than anything since then. The politically favored "solutions," pushed by activist scientists and NGOs, were the same—emissions cuts. Both produced UN conventions (the 1979 Geneva Convention on Transboundary Pollution and the 1992 UN Framework Convention on Climate Change) and protocols to implement the emissions cuts not agreed in the original conventions (the 1985 Helsinki and the 1997 Kyoto protocols, respectively).

Global warming damages are largely future tense; acid rain used the present tense, and at the height of the acid rain scare, scientists believed that the science underpinning it was stronger than for global warming. The damage was happening now. It was a slam dunk. That, at any rate, was how scientists presented the issue in their first scientific assessments on acid rain. "The facts about acid deposition are actually much clearer than in other environmental *causes célèbres*," a Canadian panel on acid rain opined in 1983.[4] The evidence appeared compelling. In North America, red spruce were dying on top of Camel's Hump in Vermont. A number of rivers in Nova Scotia had lost most of their salmon, and some fisheries in the Adirondacks were in a bad way. Lake Colden, where Teddy Roosevelt was on a fishing vacation when McKinley was assassinated, was acidic and nearly without fish.[5]

Bolin's acid rain report, produced in 1971, turned out to be a political document artfully clothed in the language of science, a further feature that would be emulated and refined with global warming. Its top-line claim brooked no uncertainty: "The emission of sulfur into the atmosphere . . . has proved to be a major environmental problem."[6] Fifty pages on, the "has proved to be" gave way to doubts and uncertainties: "It is a very difficult matter to prove that damage, such as reduced growth rates due to the acidification of the soil and related changes in the plant nutrient situation, has in fact occurred."[7] In fact, the productivity of Swedish forests had increased considerably, above all because of a more efficient system of forest management, the report said.[8] Ironically, the rapid increase in forest resources of Europe might have been due in part to acid rain, as it provides nitrate, an essential nutrient in short supply in many forest soils.[9] There was another factor Bolin hadn't considered. To improve forestry productivity, in 1960 the Swedish State Forestry Agency required pines be sourced from the Carpathians in Romania. This inland species has poor resistance to salt deposition, as happened in a big storm in 1969. As for Sweden's lakes and streams, they had shown increasing acidity which "seems mainly" caused by deposition of sulfuric acid.[10] Although fish farms routinely added lime to reduce acidity, the environmental effects of large-scale liming were not sufficiently known. "Consequently, such measures should not be taken," even though—or perhaps because—they would have undermined the case for costly emissions reductions.[11]

The report's conclusion brushed aside caveats and uncertainty to assert what had to be done. There was a "possibility" of a 10 to 15 percent reduction in timber by the end of the century. When added to possible damage to rivers and lakes, if sulfur emissions continued at their present level,

> we conclude beyond any doubt that as far as Sweden is concerned such a development cannot be accepted. A reduction in the total emissions both in Sweden and in adjacent countries is required.[12]

Bolin's conclusion was political, not scientific. It wasn't hard to make out the report's target. "The development during the last decades and, above all, the increased production of energy through fossil fuel combustion, has given rise to a series of environmental effects, which have been well-documented," it stated.[13] A 90 percent reduction in sulfur emissions would be costly. To reduce costs, "society must take measures sufficiently forcible to ensure the development of a new and less polluting technology."[14] Which one? The report didn't say, but Sweden's prime minister, Olof Palme, provided the answer in his address to the UN conference a year later. It was desirable to reduce spending on fossil fuels and convert to nuclear power, Palme said.[15]

His government had ambitious plans to build 24 nuclear power stations and develop a full-range nuclear fuel capability, from domestic uranium production to spent fuel reprocessing.*

Palme's background and Sweden's history helps explain why he was to become a leading prophet of environmentalism. Sweden had once been a European great power. Stripped of its territorial possessions, it now considered itself a *moral* power. Nuclear weapons were immoral. Neither was it good to pollute the air. Palme played the global moral role to perfection and helped keep Swedes' consciences clean. Diamond-like, Palme had a flaw. Bo Theutenberg, a Swedish diplomat who rose to be chief legal adviser to the foreign ministry before resigning in 1987 in protest at the ministry's pro-Soviet orientation, describes Palme as an "extremely intelligent" man who lacked ultimate confidence because he wasn't born into the Social Democratic Party but into a wealthy, upper-class family.[16] Leading a socialist party claiming direct descent from Marx and Engels, Palme had much to live down, and not just his family background.

As a young law student, Palme had been a fervent anticommunist. After serving as a lieutenant in the Household Cavalry, in 1947, Palme arrived at Kenyon College, in Ohio, where he majored in economics and political science. He met the actor Paul Newman and they became lifelong friends. "He spent every weekend exploring the union movement," another student recalled. Palme wrote his thesis on the United Automobile Workers union and interviewed union boss Walter Reuther. After graduating, he hitchhiked across 34 states. The trip gave him "a good picture of American society. It gave me strong feelings about social injustices," he told an interviewer in 1971.[17]

Once back in Sweden, Palme studied law and developed close links with the Intelligence Department of the Defense Staff (Underrättelseavdelningen), where his close friend Birger Elmér was working. It was the start of Palme's double life; as a politician, he would publicly criticize the West—particularly the United States—at the same time being covertly aligned with the pro-American faction within Sweden's

* The reprocessing plant, which was to be built on the Swedish west coast, attracted large protests and was eventually dropped.

governing establishment. Theutenberg was one of the few to appreciate what was going on. In addition to his diplomatic and legal career (he also became a professor of international law at Stockholm University), Theutenberg served as a reserve officer in the Swedish air force. From 1965, Theutenberg was posted at the top-secret STRIL-60 air warning system constructed with American-supplied technology to give Sweden and NATO early warning of Soviet air incursions over the Baltic. In 1976, Theutenberg was transferred to the section within the Defense Staff, heading its intelligence and espionage work, an activity in which Palme first cut his teeth. Theutenberg's *Diaries from the Foreign Ministry* provides unsurpassed evidence of the dissimulation, secrecy, and dishonesty that characterized the underlying reality of Swedish politics, none more so than of Palme walking the tightrope of his public and private positions across the deep fissure within the Social Democrats between supporters of Sweden belonging to the East or the West.[18]

As a student, Palme had provided the Intelligence Department with reports gathered from his trips in Eastern Europe. Having nothing in particular to do, in July 1953 Palme was appointed first bureau secretary at the Defense Staff's Intelligence Department. Elmér was a Social Democrat who became the first head of the extralegal IB domestic espionage organization. The IB had been established in the mid-1950s to perform "interior espionage on the Communist workers in strategic defense industries" to protect American and NATO secrets from finding their way to the Soviet Union.[19] It engaged in a kind of "opinion espionage" and operated as a secret Social Democratic body inside the Defense Staff. To those in the know, Elmér was regarded as being pro-U.S. and pro-NATO, with close ties to the CIA. In 1964, the IB was merged with the so-called T-office, which was the proper intelligence organization, under Elmér's leadership. Elmér's position as Sweden's top spy would continue until 1973, when the IB's existence was blown open by two journalists, leading to his dismissal and charges of "official lying" leveled at top Social Democrats.

Palme's position as first bureau secretary in the Intelligence Department of the Defense Staff was his first and only real job before he was appointed private secretary to Sweden's longest-serving prime minister, Tage Erlander, later that year. It was only then that he joined the Social Democrats. Thus, as a young man—he was 26 when he joined

Erlander's office—Palme had close links with Sweden's top spy as well as the prime minister, both of whom believed Sweden belonged in the West and aligned with the United States.

Four years after joining Erlander's private office, Palme entered parliament, and in 1963 he was appointed minister of communications (one of his first duties was to attend President John F. Kennedy's funeral) in charge of the state broadcasting monopoly. Two years later, he was moved to the education portfolio, but he kept control of broadcasting, thus combining state propaganda with state education. Palme was a radical education minister. He traveled extensively in Eastern Europe and reformed Sweden's schools based on ideas taken from East Germany. Children from the age of two were sent to state-run nurseries, fulfilling the ideas of the Myrdals from the 1930s. "You don't go to school to achieve anything personally," he told a group of schoolchildren, "but to learn how to function as members of a group."[20] Schools emphasized the collective; the individual was relegated. "It's useless to build up individuality, because unless people learned to adapt themselves to society, they would be unhappy," the ministry bureaucrat in charge of schools told Roland Huntford. "Liberty is *not* emphasized. Instead we talk about the freedom to give up freedom."[21]

Quick to recognize the potential of television, he set up Channel 2 of Swedish television and staffed it with leftists. The same year, the government adopted an anti-American line. Producers were told not to consider programs on the United States unless they were unfavorable. "Radio and TV became almost laughably biased, coloring news reports, and broadcasting material (some emanating from Cuba) that could only be classified as unmitigated propaganda," according to Huntford.[22] It was during the 1960s that what Theutenberg calls the "Leftist guys" were allowed to join the Foreign Ministry, which became a base for anti-U.S. and anti-NATO propaganda, steering Sweden's foreign and security policy away from the West toward Moscow.

Palme succeeded Erlander as prime minister in 1969. Although he took the Social Democrats to the left and adopted public positions seemingly contrary to those of Elmér and Erlander and to his privately held geopolitical orientation, Sweden's double game had begun before Palme. In 1950, Sweden had been the first Western state to grant diplomatic recognition to communist China. Under Palme, Sweden would do the same to North Vietnam (1970), East Germany (1972), and North

Korea (1975). In 1975, Palme would become the first Western leader to visit Cuba. A huge communist-style portrait of Palme was hung at Havana airport, where he was given a tumultuous fraternal greeting by Fidel Castro.[23]

For someone who didn't have socialism in his gut and, as Theutenberg puts it, was "learning to be a socialist," championing the environment was politically astute.[24] Environmental policy was virgin terrain over which Palme could blaze a trail without looking over his shoulder and wondering what his party comrades were thinking and what would give Sweden extra prominence on the international stage. Later on, as détente gave way to heightened East–West tension, it would provide a Third Way between the Cold War blocs and divert attention from Palme's deeply conflicted position. Environmentalism was thus a tool to bridge East–West tensions (and divert attention from them) while countering the rise of the agrarian-based Center Party, which was capitalizing on popular opposition to nuclear power.

The Stockholm conference ended with an agreed declaration "to inspire and guide the peoples of the world in the preservation and enhancement of the human environment." Acid rain was reflected in 2 of the declaration's 26 principles: Principle 21, on transboundary pollution, which declared it the responsibility of states to ensure that activities within their jurisdiction do not cause damage to the environment of other states; and Principle 22, on the development of international law on liability and compensation for transboundary pollution and other environmental damage.[25] "Transboundary" is a tracer word that tracks the extent of Sweden's impact in subsequent international agreements on acid rain.

After the Stockholm conference, Sweden used the Organisation for Economic Co-operation and Development (OECD) to push for a binding treaty to cut sulfur dioxide emissions. Then, in 1975, the Soviet Union made its first strategic use of the environment as a propaganda tool, when President Leonid Brezhnev made a speech in which he said that the environment was an issue on which East and West shared a common problem. Brezhnev's environmental push aimed to deflect Western pressure on the Soviet Union's human rights record and focus attention to other issues covered by the 1975 Helsinki Agreement on Security and

Cooperation in Europe.[26] Scandinavian countries used Brezhnev's opening to press their claims within the framework of the United Nations Economic Commission for Europe (UNECE), which included the Soviet bloc as well as the United States and Canada. The outcome was the 1979 Geneva Convention on Long-Range Transboundary Air Pollution.

Although the objective of limiting and then gradually reducing air pollution was listed as the first of the convention's objectives, like the 1992 UN Framework Convention on Climate Change, it had no targets or timetables to cut sulfur dioxide emissions.[27] The Swedes and Norwegians had proposed strict standstill and rollback clauses, but these were rejected by the United States, the United Kingdom, and West Germany.[28] Even so, a toothless agreement to monitor and evaluate long-range pollution, starting with sulfur dioxide, was a foot in the door. Sweden kept pushing. To mark the tenth anniversary of the Stockholm conference, Sweden called an international conference on the acidification of the environment. It then proposed a timetable for a 30 percent cut in sulfur dioxide emissions. A group of ten UNECE members, including the Scandinavians, Canada, Austria, and Switzerland, then formed the 30 percent club.

Meanwhile West Germany switched sides as the mainstream parties scrambled to respond to the rise of the Greens and the threat posed by the Soviet-backed Peace Movement. A November 1981 cover story in the weekly *Der Spiegel,* "The Forest Is Dying," unleashed *Waldsterben*—forest death—hysteria. It was a homegrown phenomenon promoted by Bernard Ulrich, a German soil scientist who developed a theory of soil acidification causing aluminium tree poisoning. Calmer voices were shouted down. Scientists who knew that Ulrich's contentions were based on flawed survey data were ignored. Politicians from mining regions who suggested other hypotheses were denounced as tools of the coal industry.[29]

Adroitly playing environmental politics in its campaign against NATO deployment of cruise and Pershing missiles, Moscow encouraged West Germany to host an international conference. At the conference in Munich in 1984, the Soviet bloc (with the exception of Poland) backed the 30 percent proposal, isolating the United States and the United Kingdom and burnishing the image of the Soviet Union in West Germany.[30] The following year, the Helsinki Protocol committed

Western parties to cut their emissions to 30 percent below 1980 levels and, in the case of Soviet bloc parties, a 30 percent reduction in their "transboundary fluxes" by 1993.[31] The United States and the United Kingdom did not sign the protocol.*

In response to British obstinacy, Sweden's environment minister, Birgitta Dahl, authorized a clandestine campaign against the Margaret Thatcher government. It played to an old Scandinavian cultural stereotype. In 1866, the playwright Henrik Ibsen had described how, in the future, the smoke from British coal burning would suffocate Norway's lush green vegetation.[32] In 1985, the Norwegian premier, Kåre Willoch, visited Britain and attacked Mrs. Thatcher for the "acid rain that destroys forests and fishing lakes in Norway."[33] Willoch should have given his hostess a bouquet. Thanks in part to the fertilizing effect of the nitrogen content of acid rain, later surveys showed an increase of around 25 percent in Norwegian forest growth.[34]

Acting in concert with Norway, Dahl's aim was to brand Britain the dirty man of Europe. The Swedish National Environmental Board knew that this was disinformation. Its own data showed that in 1980, East Germany was emitting more sulfur dioxide than Britain and was increasing its emissions whereas Britain's were falling.[35] But Britain was the politically favored target for Dahl, who was from the left wing of the Social Democrats and a frequent visitor to East Germany and other Soviet bloc satellites. She had chaired the Swedish Committee for Vietnam in the 1970s and defended the genocidal Pol Pot regime, dismissing reports of massacres in Cambodia as "lies and speculation."[36] Money was funneled into Friends of the Earth as part of an intensive lobbying and PR campaign. Without this money, an activist told the Swedish journal *Ny Teknik*, its campaign to portray Britain as the dirty man of Europe would never have taken off. It planned a tourist boycott of Britain by getting environmental organizations in a dozen countries to send millions of postcards: "We love your country but not your pollution." If we can reduce tourism by one percent, it will send a strong message to the government, another activist told the journal.[37] The Swedish government set up a front in London, "The Secretariat against Acid Rain," to launder around 15 million Swedish krona ($1.8m) annually to

* Neither did Poland, Spain, and Portugal.

British NGOs, though the sources of such funding weren't disclosed in Friends of the Earth's annual report and accounts—transparency for thee, but not for me, being the NGO watchword.*

They also opened up a scientific front. In 1983, the Royal Society in London announced a £5 million ($7.3m) joint research project with the Swedish and Norwegian academies of sciences. The surface water acidification program culminated in March 1990, when Margaret Thatcher welcomed her Swedish and Norwegian counterparts to a dinner at the Royal Society in London. "Let me confirm unequivocally tonight that the United Kingdom will meet the commitment which it has solemnly accepted to reduce acid emissions," Thatcher told them.[38] Hans Lundberg, the environmental secretary of the Royal Swedish Academy of Sciences, was present at the dinner. It was portrayed in the Scandinavian press as a victory even though ten times as much acid rain came from East Germany and Poland. It was environmental Realpolitik. Sweden wanted to play a role between the blocs as a "peace angel," Lundberg recalls.[39]

Across the Atlantic, Canada performed a similar role. Initially it had an ally in President Jimmy Carter. "Acid rain has caused serious environmental damage in many parts of the world including Scandinavia, Northern Europe, Japan, Canada and the Northeastern part of the United States," Carter wrote in his environmental message to Congress of August 1979. The following year, Congress passed legislation to create a ten-year National Acid Precipitation Assessment Program (NAPAP).[40] However, Carter's war against imported energy took precedence. To Canada's dismay, in February 1980, the Carter Administration launched a program to convert over 100 power stations from oil to coal.

Canada had powerful allies in Washington, in particular the backing of America's most prestigious scientific body. The National Academy of Sciences' National Research Council (NRC) published a report on acid rain in 1981. Its 13-member committee on the atmosphere and the biosphere was stacked: 3 Canadians (including chair David Schindler) and 2 Scandinavians (Svante Odén and Lars Overrein, director

* The Friends of the Earth 1986–87 annual report shows that its income more than doubled in two years, but discloses only three sources of grants, two from central and local government in Britain and the third from the World Wildlife Fund (p. 29). Why was one NGO giving money to another?

of the Norwegian Institute of Water Research). The committee members did Canada proud. Their report dismissed uncertainty; it did not mention competing hypotheses, let alone test them.

From the get-go, the NRC committee knew environmental crimes had been committed and what was to blame. Many lines of evidence showed that most kinds of atmospheric pollution had close associations with intensive human energy use, the NRC stated on page one of its report, mentioning the "postulated linkage" between increasing carbon dioxide and climatic warming (the NRC had written a report on global warming in 1977).[41] Under the heading "The Fossil Fuel Scenario: The Probability of a Crisis in the Biosphere," the NRC set out the charge sheet. "The chronicle of detrimental substances from the burning of fossil fuel could be continued for many pages." It included mercury, lead, zinc, cadmium, vanadium, arsenic, copper, and selenium, and organic micropollutants known to be carcinogens, acute toxicants, teratogens, and mutagens.[42] "Ecologists, geochemists, and climatologists are beginning to discover that in many respects man is now operating on nature's own scale, particularly through the heavy use of fossil fuels to supply the energy that runs our industrial civilization."[43]

Of all fossil fuel crimes, acid rain was the easiest to prove. "Perhaps the first well-demonstrated widespread effect of burning fossil fuel is the destruction of soft-water ecosystems by 'acid rain,'" the indictment read.[44]

> Owing to the concentrated efforts of scientists in the Northern Hemisphere, most notably in Scandinavia during the past decade, we have a much more complete knowledge of the causes and consequences of acid deposition than we have for other pollutants.[45]

The circumstantial evidence for the role of power plant emissions was "overwhelming." Many thousands of lakes had already been affected. Although sulfur and nitrogen oxides might be slightly beneficial, "the stimulus is expected to be short-lived"—an expectation confounded by subsequent forest growth abundance in Europe—but long-term precipitation "is likely to accelerate natural processes of soil leaching that lead to impoverishment in plant nutrients. When freshwater effects are considered, the positive effects are greatly outweighed by the negative," a conclusion the NRC jumped at without considering whether alternative mechanisms, such as changes in land use, might be implicated.[46]

Neither was the NRC shy in hiding its policy preference. The picture was sufficiently disturbing to merit tightening the rules on fossil fuel emissions:

> Strong measures are necessary if we are to prevent further degradation of natural ecosystems, which together support life on this planet. . . . In the long run, only decreased reliance on fossil fuel or improved control of a wide spectrum of pollutants can reduce the risk that our descendants will suffer food shortages, impaired health, and a damaged environment.[47]

Numerous studies had reported on the substantial opportunities for energy conservation and shifting to alternative energy sources. "We hope," the committee concluded, "that the report will make apparent the probable consequences of unregulated reliance on fossil fuels."[48] The NRC committee's conclusion and the means taken to arrive at it were a travesty of science. As policy advocacy, it gave Canada everything it wanted.

There were already competing explanations that better fitted the evidence. In 1977, the distinguished Norwegian geologist Ivan Rosenqvist had written a book proposing an alternative mechanism. In soils that lacked buffer capacity, where there are monocultures of pine and fir, the soil becomes more acidic. When Rosenqvist presented his findings at a conference on acid rain in Gothenburg in 1984, he was greeted by booing. The Swedish Environmental Protection Agency briefed the press that Rosenqvist was a right-wing stooge of Mrs. Thatcher and the British electricity industry. In fact, Rosenqvist was a Second World War Jewish resistance hero. He was also a communist and deemed a security risk and barred from attending scientific conferences in the United States.

Without brave scientists such as Rosenqvist, the Ronald Reagan Administration would have to fight the scientific consensus on what to do about acid rain with one hand tied behind its back. NAPAP, put in train by the Carter Administration, would independently confirm the thrust of Rosenqvist's conclusions but would deliver its final report after Reagan had left the White House. As it was, the Reagan White House would leave with the best available outcome. In the light of what subsequently became known about acid rain, it was an impressive achievement—one that was to be completely undone by Reagan's successor.

7

Acid Denial

Readers who treat the story of acid rain like a good murder mystery are on the right track. But they will find, as the mystery unravels, that most of the clues are misleading, the body is missing, and there is even doubt about whether a murder has been committed at all.

Aaron Wildavsky, 1995[1]

In 1980 the Carter Administration signed a Memorandum of Intent with Canada on transboundary air pollution. It established five acid rain work groups. When it came time to review the work groups' reports, Canada wanted peer review by the Royal Scientific Society of Canada and the National Academy of Sciences (NAS). Reagan administration officials could be forgiven for knowing what the NAS would say before being asked. Instead the White House announced that the Office of Science and Technology Policy was better qualified to integrate scientific findings with policy recommendations. In January 1982, Reagan's science adviser, George Keyworth, asked William Nierenberg, director of the Scripps Institution, to assemble a panel to conduct peer review of the work group papers.

Nierenberg's nine-member panel included the ecologist Gene Likens, who tended to alarmism about acid rain, and the University of Virginia's Fred Singer, who didn't. Meantime, pressure for action from Canada mounted. Whenever Canada pressed for 50 percent bilateral cuts in sulfur dioxide emissions, the Reagan Administration responded that more research was needed. With Pierre Trudeau as Canadian prime minister, relations between Washington and Ottawa deteriorated. In a speech to the NAS, the head of Canada's Federal Environmental Assessment and Review Office accused the Reagan

Administration of "blatant attempts" to manipulate the work groups and suppress scientific information.[2] By May 1983, EPA head William Ruckelshaus was publicly conceding ("there is no question that there is a problem").[3] His plea for modest emissions cuts was quashed by energy secretary Donald Hodel and Office of Management and Budget director David Stockman. In his January 1984 State of the Union address, Reagan proposed a doubling of funding on acid rain research. Canada sent a formal note of protest. Over 3,000 scientific studies on acid rain had been conducted, resulting in "sufficient scientific evidence . . . by prestigious scientific bodies in North America and Europe on which to initiate controls programs."[4]

Nierenberg's panel submitted its report in July 1984. In many ways, the report was a masterly compromise. It reflected Likens's obsession with microbes. It was the threat to microbes, the report said, which gave the panel its greatest concern about acid rain.*[5] The ecological problems resulting from acid rain were "sufficiently well substantiated" for the panel to recommend additional cost-effective reductions in emissions over and above those required by the 1970 Clean Air Act.[6] Yet current scientific understanding on the effects of acid rain was "highly incomplete," and the panel provided trenchant criticism of the work group reports. Work Group 1, on impact assessment, had reviewed a huge amount of data, which was often incomplete or conflicting but had failed in its fundamental task of examining the link between acid deposition and chemical and biological changes. Work Group 2, on atmospheric modeling, had greatly overemphasized the role of computer models.[7]

> The reports are not well-rounded scientific documents (e.g., they do not assess conflicting data, gaps in knowledge, strengths or weaknesses in their conclusions, or alternate theories or explanations).[8]

* In a conversation with Singer, Likens suggested valuing each microbe at one dollar. On the basis of Likens's estimate of 10,000 to 1 million microbes in a milliliter of lake water, the loss of microbes contained in 1,000 to 100,000 gallons of lake water would cost nearly $4 trillion—equivalent to the gross domestic product (GDP) of the United States at the time, illustrating why ecologists should be kept from public policy. Naomi Oreskes and Erik Conway, *Merchants of Doubt: How a Handful of Scientists Obscured the Truth on Issues from Tobacco Smoke to Global Warming*, (New York, Berlin, London, 2010), p. 90.

The provisional nature of current knowledge was highlighted by the absence of scientifically robust answers to questions responsible policymakers needed to know, the panel concluded.

The panel's deepest insight was philosophical. It criticized the general tendency to view acid rain as belonging to a socially very important class of problems that appear to be precisely soluble by a straightforward sum of existing technological and legislative fixes, a criticism that applies with still greater force to global warming. "This is deceptive," the panel warned.

> Rather, this class of problems is not permanently solved in a closed fashion, but must be treated progressively. As knowledge steadily increases, actions are taken which appear most effective and economical in the light of increasing understanding.[9]

Singer posed the killer question: Will a reduction in emissions yield proportionate reductions in acid deposition and in the environmental impacts believed to be associated with acid deposition? The other panelists didn't like letting coal-burning power plants "off the hook," so Nierenberg put Singer's question and comments into a signed appendix.[10] Singer cautioned that with emission control costs in the multibillion-dollar range, "one must question whether we are attacking a million-dollar problem with a billion-dollar solution."[11] He criticized the traditional approach of emission control of prescribing ultrastrict performance standards for new power plants (artificially prolonging the lives of less efficient and more polluting plants) as "extremely costly and wasteful to society."[12] Whereas the emissions, costs, and feasibility subgroup had not been hopeful about the efficacy and reliability of pollution controls, after talking to industry experts, Singer was optimistic the problems could be solved. Unit costs would stabilize and might even fall over time. He then proposed emissions trading on a regional scale with fully transferable emission rights. "Pollution control can then be achieved by the least-cost method," Singer wrote.[13]

The wisdom of the panel's warning and the salience of Singer's question are illustrated by what happened next. In addition to the work-group papers, the Nierenberg panel was supplied with a number of materials. More than half comprised 11 German articles, including seven written or cowritten by Bernard Ulrich plus a note by the doyen of American acid rain scientists, Ellis Cowling, on the state of German forests during late summer 1983 (the paper is dated October 7, 1983).[14] The panel was being led. Not included was a paper that turned the scientific consensus on acid rain on its head.

Two months earlier, *Science* carried an article by Edward Krug and Charles Fink, "Acid Rain on Acid Soil: A New Perspective."* The paper's focus was not on what fell from the sky but what was happening underfoot. Soil formation in humid temperate climates is an acidifying process, the paper began. Factors thought to make landscapes sensitive to acid rain are those that develop some of the most acid soils in the world. "The results of natural soil formation are those attributed to acid rain: leaching of nutrients, release of aluminium, and acidification of soil and water."[15] In Europe, several thousand years of burning, grazing, and cutting had caused severe erosion, helping to create thin soils. The temperate forests of Europe and North America were sufficiently resilient that they are now recovering. "Given the effects of vegetation on soil acidification, there is little doubt that recovery of landscapes from earlier disturbance can result in increasingly acid surface soil horizons and thickening and acidification of forest floors," the paper said.[16] The hypothesis that increased deposition of acid and sulfate—that is to say, the scientific underpinning for concern about acid rain damage to forests and lakes—was "theoretically unsound and is not supported by direct observations," Krug and Fink stated.[17]

In jumping to the conclusion that acid rain was to blame, alarmist scientists had ignored changes in land use as an alternative explanation. Recall the slam dunk evidence at the start of Chapter 6? According to Krug, damage to the lakes and streams of New England and the Adirondacks resulted from the unwinding of a period of "ecological aberration." Rain is naturally too acidic for most species of sports fish, but massive felling of trees and burning of tree stumps in the latter part of the nineteenth century had reduced the acidity of the forest floor. Soil runoff then made it possible for trout and salmon to survive. "After lumbering and burning came to an end, forests grew back, and the soil runoff, and hence the waters, returned to their natural acidity. These changes in land use often dwarf in importance the impact of acid rain."[18] Acid rain might have damaged red spruce growing in high

* Singer confirmed to the author via email (May 31, 2015) that the panel did not include Krug's paper in its review.

altitude forests in the northern Appalachians, constituting a fraction of one percent of eastern forests. Even here, the impact of acid rain was uncertain, Krug wrote. Other stress factors such as killing winters and severe drought were more important.

Krug's analysis had increasing impact on the assessments being conducted by NAPAP that Congress had authorized in 1980. In its interim report released in 1987, NAPAP said that the effects of acid rain were neither widespread nor serious and less than had been anticipated ten years before. There was sufficient uncertainty to preclude determination of the need for, or the nature of, abatement strategies such as emissions reductions.[19] The Canadians were less than pleased. Environment minister Tom McMillan attacked the NAPAP as "voodoo science."[20] The report downplayed the urgency of the problem; it was bad science and bad policy. NGOs didn't like it either. Michael Oppenheimer, an atmospheric physicist for the Environmental Defense Fund and later one of the foremost NGO cheerleaders of global warming, said the report's conclusion that acid rain contributed little to forest damage was a "startling misrepresentation." Richard Ayres of the Natural Resources Defense Council denounced the report as "political propaganda, not science."[21]

Canada's approach changed in 1984 after the Liberals were wiped out in a landslide. New premier Brian Mulroney tried sweet-talking Reagan but had no more success than Trudeau. Visiting Washington in 1988, Reagan didn't budge. Mulroney won a promise from Vice President George H. W. Bush, who pledged—before completion of NAPAP's scientific assessment—steep cuts in sulfur dioxide emissions, should he be elected. As President, Bush was as good as his word.

"The degradation caused by acid rain will stop by the end of this century," Bush declared on June 12, 1989—assuming, that is, acid rain was actually harming America's lakes and forests—when the President announced proposals to amend the Clean Air Act to cut sulfur dioxide emissions by almost half. "We touched a lot of bases as we prepared this bill, and we've had the benefit of some good thinking on the Hill."[22] Not from NAPAP, which was finalizing a report, at a cost of $570 million, that would remove the scientific justification for the new legislation.

The Clean Air Act Amendments 1990 passed by large majorities in the House of Representatives (401–21) and the Senate (89–11). The

United States then signed the 1991 Air Quality Agreement with Canada, which specifically recalls the 1979 UNECE Geneva Convention and reaffirms Principle 21 of the 1972 declaration of the Stockholm conference on the environment—a testament to Swedish soft power. In all, it had taken 19 years.

What of the science? By now, NAPAP was like an embarrassing guest whom everyone wished would just go away. NAPAP had been ready to release its final report in 1989. The timing wasn't helpful, as it conflicted with the Bush Administration's legislative push, and the draft wasn't to the EPA's liking. Even the published version shows why. Despite having its main findings spread through the document rather than summarized at the front of the report, it surgically punctured the scientific case for the steep emissions cuts required by the Clean Air Act Amendments. "The vast majority of forests in the United States and Canada are *not* affected by decline," the report stated (emphasis in the original). The partial exception was the decline of red spruce at high elevations in the northern Appalachians, for which there was some experimental evidence that acidic deposition could alter trees' resistance to winter injury.[23] There was no evidence that acid rain was responsible for regional crop reduction.[24] A particular concern of the Canadians was the impact of acid rain on sugar maple trees. Most sugar maple trees and stands were not affected by decline, though there were significant problems in Quebec and in some parts of Ontario, Vermont, and Massachusetts. Although acid deposition could not be ruled out, natural stresses including nutrient deficiencies and defoliation by insects were implicated in these declines.[25] As for lakes, NAPAP noted that liming would benefit many surface waters but only because acidic inputs are from soils or other natural sources or from non-emission anthropogenic sources such as coal mining. NAPAP backed Krug's analysis that acid deposition from power station emissions was not the cause of lake acidification.[26] As Krug pointed out, over half the acid lake capacity of the United States is in Florida, which is not downwind of large concentrations of power stations and does not receive high rates of acid rain.[27] NAPAP did find that power station emission had an impact on increased haziness and reduction of visibility. Sulfates were the dominant cause of light extinction in the eastern United States, but only one of several sources in the West.[28]

A similar downgrading had also been under way in Germany. The forest crisis ended in 1985 when the federal government announced that "the rapid increase in forest damage observed since 1982 has not altogether continued."[29] The emotive *Waldsterben*—forest death—was replaced by *Waldschäden*—forest damages. Experts from the newly created Federal Environment Ministry declared:

> There is no single type of forest damage and no single cause. We are dealing with a highly complex phenomenon which is difficult to untangle and in which air pollutants play a decisive role. [30]

This grudging nonacquittal of power station emissions came too late. Panicked by *Waldsterben* hysteria, Germany's mainstream political parties had rushed through the Large Combustion Plant Regulation to drastically cut power station emissions of sulfur and nitrogen oxides. Wishing to confer the benefits of German environmental legislation on the rest of the European Community, in December 1983 the European Commission produced a draft Large Combustion Plant Directive. The accession of Spain and Portugal into the European Community made this impractical. A watered-down version was approved in 1988, and a tougher one came into force in 2001, with the overall aim of reducing emissions of acidifying pollutants.[31]

Thus the myth that acid rain does serious environmental harm lives on in European energy policy. While the environmental benefits of reducing acid rain lie somewhere either side of zero (acid rain has a fertilizing effect), the costs are steep. In Britain, it is leading to the premature closure of 11.5 GW of coal- and oil-generating capacity, equivalent to one-fifth of peak demand. With 8.9 GW of nuclear slated for closure over the next ten years, Britain is likely to face a widening gap between its dwindling supply of low-cost electricity generation and demand.

This is not to deny the existence of acid rain caused by oxides of sulfur and nitrogen emitted from coal-fired power stations. Political scientist Aaron Wildavsky summed up the case in 1995:

> Acid rain is real; it does affect the ecosystems it falls on. . . . Both the increase in acidity and the change in acid type have had an effect on the environment. This impact was once thought by scientists and the lay public alike to be potentially destructive, but years of research have shown that devastation will not occur in lakes, forests, farms or our bodies.[32]

Media reporting helped create and inflate the acid rain scare. Wildavsky conducted a survey of articles run in seven publications (*New York Times, Los Angeles Times, Wall Street Journal, Time, Newsweek,* and *U.S. News and World Report*) and assigned them ratings from −2 (strongly critical of acid rain hypothesis) to +2 (strongly supportive). There were nearly two-thirds more proregulatory stories than balanced ones. Critical stories were outnumbered by around seven to one by ones supporting the acid rain hypothesis.

Rank	−2	−1	0	+1	+2
Percent	3%	5%	35%	39%	18%

Source: Aaron Wildavsky, *But Is It True?: A Citizen's Guide to Environmental Health and Safety Issues* (Cambridge, MA, and London, 1995), p. 300

Wildavsky noted that, for the most part, more negative articles tended to be placed farther back in the newspaper. Within articles, the number of proregulatory paragraphs dwarfed the number of antiregulatory ones.

The most egregious example was the nonreporting of NAPAP's final report. In part, this reflected the effectiveness of the EPA's media strategy to bury the good news. By the time the final report was released in September 1990, most of the legislative deals on Capitol Hill between the Bush Administration, utilities, and environmental lobbyists had been cut. NAPAP barely got any coverage. "The legislative debate was essentially over before the report came out," one reporter explained.[33] It received a one-hour hearing before a Senate subcommittee and was not formally presented to the House of Representatives. Only Senator Daniel Patrick Moynihan showed up at the hearing. As senate colleague John Glenn later told CBS, "We spend over $500 million on the most definitive study of acid precipitation that's ever been done in the history of the world anyplace and then we don't want to listen to what they say."[34]

The *New York Times*'s Philip Shabecoff, who did cover the story, told the *Washington Post*'s Howard Kurtz that NAPAP's report "should have been given more serious treatment by the media. There's a lot of good science in it." That was more than it got in the *Washington Post.* According to NAPAP director James Mahoney, the paper had "never, ever carried any article about the program, even though *The*

Post holds itself out to be the paper of record about government." Michael Weisskopf, the *Washington Post*'s environmental reporter, was on vacation when the report came out but told Kurtz that many people involved in the acid rain debate had informed him that the NAPAP report had little news value. "Just because the government threw a load of money at this thing doesn't mean it's a precious document," Weisskopf said.

> This is such a dynamic city, with so many pressure groups pushing their point of view, you don't have to do investigative reporting to find these reports. If they are truly important, they are promoted and put forward.[35]

In terms of news judgment, it was the wrong call. At the end of December 1990, CBS's *Sixty Minutes* aired a segment accusing the EPA and Congress of ignoring the NAPAP findings. Krug was the star witness: "We know the acid problem is so small that it's hard to see."[36] The EPA was furious. Hostility to Krug ran deep in the agency. In public, William G. Rosenberg, an assistant administrator at the EPA, charged that Krug had seriously misrepresented NAPAP's findings. Its report, Rosenberg said, "did show a lot of damage" from acid rain (it did not); skeptics were "interpreting the data the way they want to" (that's what Rosenberg was doing); before resorting to a line environmentalists often use when all else fails—"there are people in the tobacco industry who still argue that tobacco is not so bad," which, even if true, was irrelevant.[37] Rosenberg fired off a letter to CBS producer Don Hewitt complaining that the story was "full of half truths . . . and unfounded allegations."[38]

The EPA then prepared a PR offensive against Krug. In a rebuttal document, the EPA claimed Krug had "limited scientific credibility even in the limited area of surface water acidification."[39] After Krug had written a paper for the Department of Energy, the EPA instructed compliant scientists to review Krug's paper with a view to discrediting him. A dossier was compiled with selected quotes and sent to friendly journalists. The EPA bungled the hit job. Its fingerprints were all over press reports of Krug's "limited credibility." Krug responded that Rosenberg's attack "clearly represents libel and slander" and that "the circumstances surrounding this so-called peer review prove malice with forethought."[40] Rosenberg, a lawyer by training, immediately backed off and FedExed an apology to Krug, along with an acknowledgment that the EPA's sham peer review was not justified. "We respect Ed Krug as a scientist," Rosenberg's spokesman told *Reason* magazine. "We simply disagreed with his scientific conclusions."[41] Even this was untrue. In private, Rosenberg and the EPA admitted that the acid rain story was pulp. According to columnist Warren Brookes, after the *Detroit News* had first started running his stories on the NAPAP study, Rosenberg told the paper's editorial board that, essentially, "we know acid rain is not that big a deal when it comes to forests and lakes. The big issue is health effects."[42]

The shift in justification from environmental to public health benefits is confirmed by a 1997 paper, *The Costs and Benefits of Reducing Acid Rain,* produced by the Resources for the Future (RFF) think tank. The paper's title is somewhat misleading, as the putative benefits identified in the paper do not come from reductions in acid rain, but from reductions in acid rain's chemical precursors, principally sulfur and nitrogen oxides and sulfate particulates. The RFF paper notes that, in contrast to the focus of attention in the 1980s, effects on forests and aquatic systems had still not been modeled comprehensively "because of a lack of proper scientific and/or data and models."[43] Overall, RFF estimated average per capita health benefits of $62.79; $9.15 of benefits from improved visibility and $0.62 of aquatic benefits from the avoided costs of putting lime into lakes compared to per capita costs of $5.30.*[44]

Retrospecctive justification shifting not only damages scientific integrity; it results in bad policy. If public health had been the policy objective, emissions trading was the wrong answer. Acid rain is regional—emissions from power stations in the Midwest cause rain that is more acidic to fall on the Northeast. By contrast, putative public health harm is highly localized, being dependent on specific local factors such as population density, the level of non–power station sources of sulfur dioxide, and local climate. Locational heterogeneity of emissions sources implies that a uniform damage function and, hence, market price is meaningless. If the public health benefits from cutting power station emissions are believed, a national cap-

* It appears that the RFF researchers had not taken on board the NAPAP finding that cutting emissions would make little difference to lake acidity.

and-trade system is not the solution. Rather, the pragmatic justification for cap-and-trade is recognizing that however weak the scientific and economic case might be, politicians and regulators were determined to "do something" and that the EPA is not competent to fashion tailor-made, local regulations. Whatever is done—however ineffective—should be done at least cost. Incorporation of cap-and-trade in the Clean Air Act Amendments gives the United States the lowest-cost route to meet an arbitrary emissions reduction target, costing around half initial estimates, but means the quantified health benefits are essentially fictitious.

The shift from environmental to public health harm certainly made it easier for the EPA to fabricate flattering benefit:cost numbers, a shift that would be replicated by the Obama Administration to justify its war on coal. Health concerns are better at winning voters' support than environmental ones, and it is not too hard to sell the idea that what comes out of the chimneys of coal-fired power stations is injurious to people's health. But the evidence of harm to human health is thin, to say the least. A 2003 study on particulate matter in outdoor air pointed out that mortality risk estimates from epidemiologic studies are weak and derived from studies unable to control for relevant confounding causes. No form of ambient particulate matter—other than viruses, bacteria, and biochemical antigens—has been shown, experimentally or clinically, to cause disease or death at concentrations remotely close to U.S. ambient levels.[45]

A 2009 analysis of monthly hospital admissions for lung diagnoses and air pollution levels for five different pollutants (carbon monoxide, particulates, sulfur dioxide, nitrogen dioxide, and ozone) across eleven Canadian cities from 1974 to 1994 found the health effect of air pollution to be numerically "very small" and in almost all cases either insignificant or negative (i.e., the pollutants are beneficial to health) but did find consistent evidence that lower tobacco smoking rates led to fewer hospital admissions and shorter stays. Reviewing the literature on previous studies, the authors found a wide range of contradictory results, which they ascribed to model uncertainty and the dependence of model calibration estimates that were "inherently opportunistic."[46] The authors suggest the possible existence of a threshold effect above which there are adverse effects on health from air pollution, citing heavily polluted cities in the developing world where pollution levels can be well over double the maximum levels observed in Canadian cities.

In 1999, Gene Likens was writing, "Acid rain still exists and its ecological effects have not gone away."[47] Few people cared. The world had moved on. Unlike global warming, alarm about acid rain is history. We can therefore view its complete life cycle. Acid rain had not been "solved." It faded away. Retrofitting power stations to cut emissions of sulfur and nitrogen oxides, fitting catalytic converters on vehicles, did not create trillion-dollar rent-seeking opportunities remotely comparable to what global warming has with renewable energy. Environmental scares relying on the present tense run the risk of present-tense falsification. Scientists claimed trees were being damaged and forests would die. The evidence showed that they weren't. By contrast, evidence of damage caused by predicted catastrophic global warming is, by its very nature, speculative, with predicted global temperatures up to 4.8 degrees centigrade (8.6 degrees Fahrenheit) higher by century end than the average for the second half of the nineteenth century.[48] Climate scientists are notably reluctant to nail their colors to the mast and say by when observed global temperatures should catch up with computer model projections before they change their dire prognoses. Global warming's predictive elasticity and its avoidance of an empirical falsifiability test give it more staying power than acid rain. We can also see that regulatory efforts to curb acid rain were political posturing. At no point did any environmental regulator or any of their political masters acknowledge that the scientific basis for policies against acid rain had gone. Ignorance prevails and history is distorted. In their 2010 treatment of acid rain, *Merchants of Doubt,* authors Naomi Oreskes and Erik Conway deal with NAPAP by the simple expedient of not mentioning the findings contained in NAPAP's final report.

Acid rain was not just a scare. It was a scientific scandal. "Acid rain is a serious environmental problem that affects large parts of the United States and Canada," the EPA's website still falsely proclaimed at the time of writing.[49] To get legislation to cut power station emissions through Congress, the EPA knowingly suppressed evidence. It then trashed the reputation of a scientist whose only crime was concern for the truth.

With global warming, the public is asked to put their trust in the judgment of national science academies. Acid rain provides an objective test of their trustworthiness. The national academies of five countries—those of America, Canada, Britain, Sweden, and Norway—produced reports on acid rain that were biased and unscientific. Their errors remain uncorrected and unretracted. Collectively, the academies stand guilty of collusion in scientific malpractice. Their authors presumed to know too much and downplayed uncertainty and lack of knowledge in furtherance of a political agenda. "Science," the philosopher Karl Popper wrote in 1960, "is one of the very few human activities—perhaps the only one—in which errors are systematically criticized and fairly often, in time, corrected."[50] By Popper's standard, what they practiced was not science. It was political advocacy. It would be a template for what was to come.

8

Double Cross

The arms race is accelerating. The development of new nuclear weapons seems to suggest that the nuclear powers may actually consider fighting a nuclear war.

Olof Palme[1]

Pacificism . . . is in the West; the missiles are in the East.

François Mitterrand[2]

Leonid Brezhnev's acid rain intervention in 1975 (Chapter 6) illustrates how the development of environmentalism became entwined in the politics of the Cold War. In West Germany, environmentalism developed sideways and upward from the grass roots. By the late 1970s, the Sixty-Eighter student activists were washed up, living on the margins of society. They were no longer young and were going through something of a midlife crisis. Isolated as a result of revulsion against the Far Left terrorism that had swept West Germany in the mid-1970s, they viewed environmentalism as a way back to relevance, and the prospect of acquiring power and well-paying jobs played out against the backdrop of the most serious attempt to undermine the Atlantic Alliance in its history. In a decade of danger, faced with mass demonstrations and the pull of a neutralist-tilting left, West Germany's loyalty to the West remained rock solid.

By contrast, Sweden deployed environmentalism as a top-down tool of social control. Similarly, state-sponsored anti-Americanism was used as a social safety valve. As with acid rain, Sweden would be the first to make global warming a political issue, and Palme had been the first politician to meld environmentalism and disarmament when, in 1972, he used his speech at the Stockholm UN environment conference to attack the United States over the Vietnam War in language that could have been spoken by Herbert Marcuse. "The immense destruction brought about by indiscriminate bombing, by large scale use of bulldozers and herbicides is an outrage sometimes described as ecoside," Palme declared.

> We know that work for disarmament and peace must be viewed in a long perspective. It is of paramount importance, however, that ecological warfare cease immediately.[3]

In an address broadcast on Swedish radio six months later, which became known as the Christmas Speech, Palme denounced the bombing of Hanoi.

> We should call things by their proper names. . . . Many such atrocities have been perpetrated in recent history. They are often associated with a name: Guernica, Oradour, Babi Yar, Katyn, Lidice, Sharpeville, Treblinka. Violence triumphed. But posterity has condemned the perpetrators. Now a new name will be added to the list: Hanoi, Christmas 1972.[4]

Days after the announcement of the Paris Peace Accords in January 1973, Palme told the Swedish parliament, "The guilt of western civilization is great."[5] Did Palme think of himself as belonging to the West?

> I don't see anything special about Florence, or Paris or Rome. I feel more at home in Prague and Warsaw and Sofia. . . . Western culture? What does it mean to us?[6]

The Christmas Speech unleashed a torrent of criticism in America and a near breaking off of diplomatic relations.* To smooth things over, Palme wrote to President Richard Nixon to say how personally grateful he was to the U.S. for his year of study there and how much he admired American history and ideals. He also wrote to a friend from his student days in America. "I am deeply worried, disgusted and almost desperate because of the incredible folly of the Vietnam War. . . .

* When the American ambassador in Stockholm returned to the U.S., he was not replaced and the U.S. requested his Swedish counterpart in Washington not be replaced either. Full diplomatic relations were not restored until 1974.

For somebody who loves America, her people and her institutions, this is particularly tragic."[7] That affection, rarely expressed in public, was subordinated to politics: State-sponsored anti-Americanism was a technocratic tool to stabilize Swedish society by coopting the student protest movements sweeping across the West in the late 1960s. "Where Western governments fought the trend, the rulers of Sweden made an ally of it," Huntford wrote in 1971.[8] Anti-Americanism had a cathartic quality akin to the two-minute hate in Orwell's *1984*, Huntford thought. "I feel so *emancipated*," a Swedish housewife said in a newspaper interview after a particularly violent demonstration outside the American embassy.[9] A director of Bofors, a major Swedish arms exporter, told an American journalist, "I can only say thank God for all this anti-Americanism and Vietnam protest. . . . I'm very relieved that anti-Americanism has kept the heat off us. It's probably the only thing that could've done."[10]

Sweden's official anti-Americanism could have caused problems for Palme, given his intelligence background and links to the pro-American spy chief Birger Elmér. It also served as a decoy for Sweden's secret security understanding with the United States and NATO, which, if known, would have implied a public repudiation of Sweden's fourteen-decade policy of neutrality. The so-called 1812 Policy had been the guiding principle of Sweden's foreign policy since the Napoleonic Wars and had helped Sweden become one of the most prosperous nations in the world in the first half of the twentieth century. By the early 1950s, the prime minister, Tage Erlander, had concluded that it would be impossible to keep Sweden without friends and allies. At first he wanted to be able to buy sensitive and advanced military equipment from the United States and NATO. The arrangement deepened, as he then wanted Sweden to be able to count on military cooperation and assistance if attacked by the Soviet Union. However, Erlander's foreign minister, Östen Undén, was ideologically more pro-Soviet and wished to avoid at all costs anything that might provoke Joseph Stalin.*

* Undén's concern to avoid anything that might antagonize the Soviet Union extended to the disappearance and murder of Raoul Wallenberg, who had saved thousands of Hungarian Jews from Nazi death camps. Undén therefore rejected a deal to free Wallenberg. By contrast, the Swiss, who were offered a similar deal, worked hard to free two of their diplomats kidnapped by the Soviets. Source: Reuters, "Sweden Rejected

Without informing Undén, Erlander went ahead and developed a long-term understanding with Washington, which included secret arms purchase agreements and an undertaking, also secret, to jointly defend Sweden in the event it was attacked. Knowledge of the full scope of the understanding was restricted to an extremely limited number of individuals. Military personnel knew a little piece here, a little piece there. Thanks to Stig Wennerström, a Soviet spy and a colonel in the air force, the Soviet Union learned early on about Sweden's decision to align with NATO. The only people kept completely in the dark were the Swedish people. To this day, no state document has been produced setting out exactly what the understanding entailed.* Military cooperation was deep and extensive. Runways were lengthened to accommodate NATO reinforcements; a military headquarters-in-exile was prepared; NATO submarines could use Swedish waters; signals intelligence was shared. Ostensibly to bolster its "armed neutrality," Sweden was given the latest American technology to build the fourth-biggest air force in Europe.

In such a political culture, it would be foolish to take anything at face value. Despite Palme's official and highly vocal anti-Americanism, he was still very keen to uphold the "special military links to the U.S." and even count on American "military help" should Sweden be attacked by the Soviet Union. Even during the "frosty years 1972–1974" in U.S.-Swedish relations, following his Christmas Speech, Palme sent signals to Washington that his criticism should not affect the "traditional military pattern between U.S. and Sweden."†[11] And it really never did! In his diaries, Theutenberg states his belief that it was secretly maintained throughout Palme's years as prime minister and possibly even after that.[12]

Swap for Wallenberg," January 4, 2001. MIT professor of political science Susanne Berger has also extensively investigated the Wallenberg case. See, for example, Susanne Berger, *An Inquiry Steered from the Top?*, March 1, 2015.

* A 2014 official inquiry described Sweden's cooperation with NATO half accurately and half misleadingly as "largely informal and partly secret." The absence of documentation can be seen by the inquiry's statement that in the event that Sweden was attacked, there were "substantial indications that for much of the Cold War, a preparedness existed in the West to contribute to Sweden's defense." Government Offices of Sweden, *International Defence Cooperation Efficiency, Solidarity, Sovereignty: Report from the Inquiry on Sweden's International Defence Cooperation* (Stockholm, 2014), pp. 27–28.

† Palme had a strong personal rapport with Henry Kissinger.

Palme's playing to the leftist/anti-American gallery worsened Sweden's strategic dilemma. On becoming prime minister in 1969, he had cut a deal with the Left Party–Communist MPs to halve the defense budget by 1972. It was a dangerous game. Weakening Sweden's defenses made Sweden more vulnerable to the Soviet Union at the same time as making it more dependent on its secret alliance with the United States. By the end of the decade, there was growing evidence of Soviet submarine movements in Swedish waters. In 1981, one of them, U-137, ran aground on Sweden's southern coast near the Karlskrona naval base, inspiring the Tom Clancy thriller *The Hunt for the Red October.** Palme's opponent, Thorbjörn Fälldin, was prime minister at the time, and Theutenberg, the top legal adviser at the foreign ministry, had no difficulties in formulating a strong note of protest that was delivered to the Soviets.

Less than a year later, there was an incident in the Hårsfjärden fjord near Stockholm, close to two naval bases, where the Swedish navy trapped another foreign submarine. It occurred after elections in September, just as Palme was forming a new government that took office on October 8. The incident was a huge embarrassment to the new government. The last thing it wanted was to start its term with a big diplomatic dispute with the Soviet Union, especially as Palme had just spent the previous two years heading a high-profile international commission on disarmament. *Common Security: A Programme for Disarmament* had been published just before the election. "Common security" was the way Moscow framed its propaganda offensive against NATO rearmament. Palme's report was replete with Soviet ideas such as nuclear-free zones designed to undermine NATO's response to Moscow's SS-20 intermediate-range nuclear missiles. The general mantra from the Swedish left was that there was not sufficient proof of submarine activities, and even if there were, they were NATO, not Soviet, submarines. Instead of issuing a robust protest as Fälldin had done, Palme formed a commission to investigate the violation of Swedish territorial waters by foreign submarines. Nonetheless, in April 1983 the commission reported that indeed the Soviet Union was respon-

* U-137 was really called S 363 and armed with nuclear weapons. According to Commodore N-O Jansson *Omöjlig Ubåt* ("Impossible Submarine") (Gothenburg, 2014), its mission was probably to pick up a Spetsnaz force that had been landed some days previously near Karlskrona (p. 173–183).

sible for the Hårsfjärden submarine incident. Palme himself then handed the Soviet ambassador a protest note that Theutenberg had drafted. This was a rare occasion when Palme overrode what an exasperated Theutenberg in his diary called that "damned entourage" of "KGB/IA [Internationella Avdelningen, Swedish for the International Department of the Central Committee of the Soviet Communist Party]/STASI/GRU-influenced people" that forced Palme to "watch his tongue." It was an essential part of the "great paradox" of Olof Palme; he wanted to do something but acted in a totally different direction, Theutenberg recorded.[13]

This gang of "Leftist/Socialist guys," who became known as "Palme's boys," secured top bureaucratic positions in the foreign ministry. When Palme became prime minister again in 1982, Pierre Schori, the international secretary of the Social Democrats, was immediately given the top permanent post at the foreign ministry. Schori came from the revolutionary Far Left and was a pupil of Fidel Castro, whom he had met on several occasions, and was influenced by Che Guevara.[14] Even when the Social Democrats were out of power, the party decided who would be Sweden's ambassador to the UN. From 1977 on, Stockholm sent a string of Palme's boys to Turtle Bay to head Sweden's UN mission, starting with Schori's predecessor as the Social Democrats' international secretary, Anders Thunberg (1977–1982), followed by Anders Ferm (who had been secretary to the Palme commission on disarmament), and, later, by Schori himself (2000–2004). From its perch at the UN, this Far Left network was deeply involved in coordinating not only Sweden's disarmament policy but also its diplomatic offensive on the environment across the UN and its agencies, especially UNECE (for acid rain) and the United Nations Environment Programme (UNEP), which had been established in the wake of the Stockholm environment conference. UNEP's first executive director was Maurice Strong, succeeded in 1975 by Mustafa Tolba, who had headed Egypt's delegation to the Stockholm conference. UNEP was also part of the Palme boys' network. Ulf Svensson, a Foreign Ministry diplomat who was one of the leftist gang, served as director of UNEP's regional office for Europe, where he pursued a disarmament and environmental agenda, commissioning reports on how the arms race purportedly was damaging the environment.

Surrounded by a pro-Soviet entourage at the top of the Foreign Ministry, Palme aroused suspicion that he was a KGB agent. Professionals saw him otherwise. To them, he was, in Theutenberg's words, a "pro-American, pro-CIA-guy," heavily influenced by the pro-CIA spy chief Birger Elmér.[15] Playing a double game, Palme campaigned against deployment of Pershings and cruise missiles and urged Denmark and Norway to leave NATO in favor of a neutral, nuclear-free Scandinavia.[16] A former Swedish ambassador to Moscow remarked that in the field of disarmament in the 1980s, Sweden's foreign policy was unusually parallel to Soviet positions.[17] Palme was trapped by public adulation from the international left and by the coterie of pro-Soviet officials running the Foreign Ministry, a position he found himself in at the time of his death in 1986 and might well have led to it.

Sweden's secret alliance with NATO, its manufactured anti-Americanism, and the double game played by Palme illustrate the depths of deception practiced by the Swedish Deep State. Things are not what they appear to be; in fact, they can be the opposite of what they seem. In such a climate of secrecy and deceit, it would be naïve to take Palme's environmentalism as merely a matter of belief or political principle, as genuine attempts to solve genuine problems. Rather it was fashioned and deployed for reasons of political utility, reasons which, in the nature of the exercise, have to be inferred, as the protagonists were not in the business of giving the game away. Being a prophet of environmentalism had a number of advantages for Palme: to himself, as the leader of a party in which he was something of a social transplant and to Sweden, enabling it to display its virtue to the world and project Scandinavian soft power far above Sweden's rank. In a period of geopolitical crisis, environmentalism provided the international left with a project to anesthetize East–West tensions, something Moscow had been trying to do since Leonid Brezhnev's 1975 environment speech. Sweden's apparent neutrality gave it an elevated moral status as a voice speaking for the security and health of the planet. This elevated status was especially attractive to the enigma that was Olof Palme, the secret pro-American surrounded by pro-Soviet Foreign Ministry advisers, who was now being hailed by the international left as a world historic figure.

In this game, science was the servant of the interests of the state. It was a task Palme's friend and science adviser, Bert Bolin, accepted with alacrity when, as we saw, he supervised the Swedish government report on acid rain. It was a function that would be reprised with the nuclear winter scare of the early 1980s. Bolin would play his most important lead role at the end of the 1980s with global warming, when Sweden would take the lead in carrying out the groundwork for the creation of the IPCC. Swedish soft power was not hard enough to prevent the Soviet Union from violating its waters and subverting its foreign policy, but it could pack sufficient punch to help put global warming on the international agenda.

9

Born Again Greens

The movement for survival (the peace movement) grew quite naturally out of the movement for life (the ecology movement).

Werner Hülsberg[1]

. . . the Greens in West Germany wanted to take the left-wing concepts from the past and, item by item, recycle them into notions suitable for the modern age. Instead of the old proletarian metaphysic with its catastrophic vision of capitalism and its dream of a future proletarian society, the Greens proposed a new ecological metaphysic with its own catastrophic vision of capitalism and its dream of a new ecological utopia. Instead of the cult of the factory, the cult of the forest. Instead of the class war, the ecological struggle. Instead of the socialist millennium, the ecological millennium. Instead of the color red, the color green.

Paul Berman[2]

Sweden was a strategic flank in the Cold War; West Germany was the central front and throughout remained a stalwart member of the Atlantic Alliance. Residual support for communism evaporated after the establishment of East Germany, and in 1956 the Communist Party of Germany (KPD) was banned.* Three years later, the Social Democratic Party of Germany (SPD) adopted the pluralist, anti-Marxist Bad Godesberg Program and supported "a free market wherever free

* With the help of the East German communists, in 1968 the KPD was reconstituted as the DKP.

competition really exists."[3] For two parties with the same name, the German and Swedish social democrats could hardly be more different. As Werner Hülsberg, a prominent Green activist and eco-socialist, lamented, the SPD had replaced class struggle with the American way of life.[4] Worse still from an environmentalist perspective, the Bad Godesberg's preamble championed man's peaceful use of atomic power to make human life easier and "create prosperity for all if he uses his ever growing power over the forces of nature solely for peaceful ends."[5] Bad Godesberg was, in the words of an early eco-socialist and now leading critic of global warming, Fritz Vahrenholt, a "hymn to nuclear power."[6]

During West Germany's first three decades, environmentalism had mostly been a preserve of a tiny fringe on the Far Right. There were a few politicians within the mainstream parties concerned about the environment, but the mainstream as a whole wasn't. Before the late 1970s, according to the British historian Anna Bramwell, "Greenness was seen as an incipiently sinister conservative or even Fascist idea in German thought."[7] At the other end of the political spectrum, in 1960, the disenfranchized Far Left formed the Extra-Parliamentary Opposition (APO), emulating the British Campaign for Nuclear Disarmament. Radicals took over the SDS (same initials as the American Students for a Democratic Society but standing for the Sozialistische Deutsche Studentenbund), the SPD's youth wing. A student counterculture flourished in West Berlin and Frankfurt, where prominent figures of the future Greens, notably Joschka Fischer and Daniel Cohn-Bendit, were active. The Frankfurt School was a major influence, and in 1966 Marcuse visited Frankfurt and met with the SDS there.

There were large demonstrations against the Vietnam War. On June 2, 1967, a demonstration in Berlin turned deadly. A policeman shot and killed a student. A photograph of a woman cradling the student's head became an iconic image in West Germany. For the *New York Times* four decades later, it was the shot that put conservative West Germany on course to evolve "into the progressive country it has become today." To the students and the Far Left, the shooting revealed the fascistic nature of West Germany. Only in 2009 was it revealed that the West Berlin cop was a Stasi agent and member of the East German communist party. "I would never, never, ever have thought that this could be true," Stefan Aust, a former *Der Spiegel* editor and author of a sympathetic book on Far Left terrorism, commented.[8]

Three days after the shooting, Egypt, Jordan, and Syria attacked Israel. With the Six-Day War, the New Left "discovered" the Palestinians, as historian Paul Hockenos puts it, opening their eyes to "the fact that Israel's staunchest allies just happened to be Washington, Bonn, and the right-wing Springer media group."[9] As Paul Berman wrote in a brilliant and subtle 2001 essay first published in *The New Republic*, through a tortured inversion (a particular speciality of the New Left) Israel became the imperialist aggressor and the Palestinians the anti-Nazi resistance, whereby

> the New Left's vision of a lingering Nazism of modern life was suddenly re-configured, with Israel in a leading role. Israel became the crypto-Nazi site par excellence, the purest of all examples of how Nazism had never been defeated but had instead lingered into the present in ever more cagey forms. What better disguise could Nazism assume than a Jewish state?[10]

Deeds followed words. In the final weeks of 1969, the 21-year-old Fischer formed part of a five-person SDS delegation to a Palestine Liberation Organization (PLO) solidarity congress in Algiers. Years later, photos surfaced showing Fischer in a dark turtleneck sweater and jacket staring thoughtfully into the camera with a translation headset on his head. "The rebel students' insensitivity to Jewish issues and unmediated pro-Arab leanings were among its glaring blind spots," Hockenos observes in his mostly sympathetic account.[11]

A May 1969 survey found that 30 percent of West Germany's high school and university students claimed to sympathize with Marxism or communism.[12] According to the communist commentator Gerd Koenen, the early seventies were even more left than the late sixties.[13] "Though ultimately just a fraction of the greater German left," Hockenos finds, "the sheer number of those who went the way of radicalism in the 1970s was astonishing. In West Germany, many more took this path than [they] did in France or other countries, including the United States."[14] For Koenen, 1967–1977 was the Red Decade. It certainly was a decade of violence and turmoil, which would end only with the complete defeat of terrorism and the internal exile of the New Left. Yet within three years, they would be on the march again in a new guise.

The SPD under Willy Brandt won a huge election victory in 1972. Voters liked prosperity and Brandt's *Ostpolitik*—the policy of normalizing relations between the two Germanys. Within two years, the atmosphere of optimism had gone, as had Brandt from the chancellorship. The Stasi had penetrated Brandt's office. By the early 1970s, its agent, Günter Guillaume, who had "escaped" from East Germany in 1956, was attending all meetings of the SPD and parliamentary leadership. Early one morning in April 1974, Guillaume and his wife were arrested in their Bonn apartment. Dressed only in a bathrobe, Guillaume blurted: "I am an officer of the [East German] National People's Army." According to interior minister Hans Dietrich Genscher, "it was basically only Guillaume's own declaration which convicted him."[15] Two weeks later, Brandt resigned, though he remained SPD leader. Markus Wolf, the East German spymaster, subsequently admitted that putting an agent in Brandt's office had helped destroy the career of "the most far-sighted of modern German statesmen."[16] Wolf was right. In Helmut Schmidt, Brandt's successor, the West gained its most impressive leader of the 1970s.

The 1973 oil price shock ended West Germany's postwar economic miracle and seemed to validate the prophecies of the environmentalists. Despite the deepening political and economic challenges facing West Germany, the realization dawned among Marxist intellectuals on both sides of the Iron Curtain that communism had lost the economic race against capitalism, so the rules of the game had to be redefined. The Club of Rome's *The Limits to Growth* became a bestseller, as did the even more darkly pessimistic *The Plundering of a Planet*, a homegrown effort by the Christian Democrat and future Green Herbert Gruhl. On the other side of the Iron Curtain and across a rapidly narrowing ideological divide with the anticapitalist Far Right, Marxist intellectuals such as Wolfgang Harich and Rudolf Bahro were arguing that economic growth was threatening the planet. In 1975, Harich wrote *Communism without Growth*, which clothed Club of Rome eco-doom in Marxist dialectics. Bahro made a similar journey in *From Red to Green* (the title of a 1984 set of interviews Bahro gave to the *New Left Review*). Marxists should take their inspiration from Mao Zedong rather than Stalin. "I thought Mao could steer a political course that bypassed Stalinism," Bahro explained.[17] "Mao's point of departure was to ask whether communism can be built in more backward conditions"—as if conditions in West Germany in the 1970s were more backward than in Russia in 1917.[18]

Both Harich and Bahro were imprisoned and later left East Germany. While in the West, Harich made contact with Gruhl before returning to East Germany in 1981. Although Bahro stayed in the West to become prominent in the Greens, he did not repudiate his allegiance to East Germany.

> My general position on the GDR [German Democratic Republic] . . . is basically shared by most of the Party membership. We are all against capitalism, and we have achieved something in the GDR that works, even if it doesn't always work all that well.[19]

A profession of loyalty such as this should be taken for what it was. East Germany's Socialist Unity Party (SED) was one of the Soviet bloc's nastiest communist parties. Its effect on West Germany was pervasive and malign. In noting the influence of the SED on the West German New Left, Berman comments that the SED "expressed a fiercer hatred for Israel than did any other ruling party in Europe" and provided aid to the Ba'ath party in Syria and Iraq.[20]

The West German New Left was also saturated in the doctrines of the Frankfurt School. For a time, Cohn-Bendit ran one of the large number of "anti-authoritarian" kindergartens that sprang up across West Germany, "the essays of Adorno ornamenting every anti-authoritarian classroom."[21] Cohn-Bendit also urged Fischer to join his small cell, the Revolutionary Struggle, modeled on the Italian Lotta Continua and its aim of radicalizing the proletariat. "What made Fischer so special in this milieu was thuggish elements like Hans-Joachim Klein respected him. They could communicate with fellow-prole Fischer in a way that they never could with the cerebral student revolutionaries," Hockenos writes.[22]

Involvement in alternative education, Berman notes, had been a short step for SDS leader Bill Ayers to the bombings of the Weather Underground in America at the end of the 1960s. A number of Ayers's West German counterparts made a similar journey. Having studied the canonical texts of the Frankfurt School, West German radicals viewed Western culture as an instrument of oppression: "The terrorist logic, such as it was, drew on Marcusean social criticism: the criticism that saw no hope at all in Western society."[23] Their answer was the bomb and the bullet. After training at a PLO camp in Jordan, the Red Army Faction (RAF), as the Baader Meinhof gang were first known, and the Second of June Movement (after the 1967 demonstration in West Berlin) carried out a wave of killings and kidnappings. In early May 1972, RAF terrorists set off bombs at U.S. military bases across West Germany. One U.S. officer was killed and thirteen injured. "You must know that your crimes against the Vietnamese people have made you new, bitter enemies and that there won't be any place in the world where you can be safe from revolutionary guerrilla units," read a hand-typed justification for the bombings.[24] Others such as the Revolutionary Cells, which Klein joined, collaborated with the Popular Front for the Liberation of Palestine (PFLP). At his terrorist training camp, Klein discovered "European leftists singing left-wing songs . . . and, in another part, European fascists singing fascist songs," Berman records.[25] In December 1975, Klein and other PFLP terrorists, led by the Venezuelan terrorist Carlos the Jackal, attacked and kidnapped OPEC oil ministers in Vienna.

Five months later, Ulrike Meinhof was found hanged in her prison cell. The discovery led to street violence. In a police sweep, several Far Left radicals were arrested. Fischer spent the night in a police cell. According to Hockenos, it was something of an epiphany—Fischer's "tryst with violence" had gone too far. On the evening of his release, Fischer met Cohn-Bendit and other leaders of their group and agreed not only to turn away from violence but also to persuade others on the Far Left to do so as well. There was disagreement about who should take the lead. "Some wanted me to do it," Cohn-Bendit recalled, "but I insisted that it had to be someone from the militant Left, like Joshka."[26] Thus Fischer and his circle began their climb out of the pit of New Left nihilism. They didn't stop the violence. Worse was to come.

On June 27, 1976, an Air France flight from Tel Aviv to Paris was hijacked shortly after taking off from a scheduled stopover at Athens. It was a combined Revolutionary Cells–PFLP operation. The four terrorists were led by Wilfried Böse, an intimate of Fischer's Frankfurt circle. By prearrangement with Ugandan dictator Idi Amin, the plane was flown to Entebbe. The hostages were transferred to a disused passenger terminal. There, Israelis and Jews were selected and then separated from the rest, most of whom were released. The Israelis were to be executed unless the hijackers' demands were met. On the hijacking's seventh night, Israeli commandos flew to Entebbe and killed Böse and the other hijackers for the loss of three of the hostages and their commander, Yonatan Netanyahu, elder brother of the future

Israeli prime minister. It was the terrorists' first major defeat. Even so, the killings continued, to culminate in the "German Autumn" of October 1977.

On September 5 of that year, RAF terrorists kidnapped the head of the German Employers' Association, Hanns Martin Schleyer, and demanded the release of their imprisoned comrades. Schmidt stood firm. "In the long term, terrorism has no chance. Not only is the will of the state against it, so is the will of the people," Schmidt told West Germans.[27] Five weeks later, four PFLP terrorists hijacked a Lufthansa flight from Mallorca and demanded the release of imprisoned Red Army Faction leaders (the future Green Otto Schily was one of their defense attorneys). After being refueled in Rome, the plane flew around the eastern Mediterranean and the Gulf before landing in Mogadishu. Schmidt despatched an elite team of German police to Somalia, where they stormed the plane, shot the hijackers, and freed all the passengers. That night, three jailed terrorists committed suicide (a fourth tried and failed). The next day, Schleyer's body was found. He had been shot. It was a gruesome end to a violent, bloody spasm that had claimed around 50 lives but had left West Germany's democratic values intact. The vast majority of West Germans were disgusted by the terrorism. As police conducted a massive operation to track down the terrorists, the Sixty-eighter radicals were completely isolated.

The New Left's external crisis was mirrored by an internal one: What was the New Left actually *for*? Along standard Frankfurt School lines about the fascistic nature of Western society, in Berman's telling, the New Left was motivated by a fear that

> Nazism had not been defeated after all—a fear that Nazism, by mutating, had continued to thrive into the nineteen-fifties and sixties and onward, always in new disguises. It was a fear that Nazism had grown into a modern system of industrial rationality geared to irrational goals.[28]

New Left terrorism, especially that specifically directed at Jews, together with its support for the communist regimes of South East Asia, triggered an existential crisis.

> For it was suddenly obvious to anyone with eyes that huge portions of the New Left had ended up supporting a cause that, in

> the case of the Palestinian guerrillas and their allies in Germany and other countries, was on a tiny scale resurrecting the old manias of the Nazis of the nineteen-thirties and forties, and, in the case of the Cambodian Communists, was engaged in slaughter by the millions. Anti-Semitism and genocide, a familiar twosome. And it became obvious that the New Left in its more radical or revolutionary version was not, as everyone had imagined, an anti-Nazi movement. On the contrary.[29]

Why hadn't the New Left's fear of Nazism prevented it from ending up Nazi-like? The New Left shared the nihilism and antihumanism of the Nazis, which made it highly susceptible to adopting similar positions as the Nazis—and not just once, with its anti-Semitic violence, but a second time, when its leaders became born again Greens.

Inadvertently, Schmidt provided the radicals with a path to respectability and political ascendancy. In response to the oil crisis, the SPD decided to build more nuclear power stations. It sparked a grassroots reaction and the spontaneous formation of local Citizens' Initiative environmental groups. In a 1977 *Der Spiegel* opinion poll, 43 percent opposed any further construction of nuclear power stations, with 53 percent in favor. The *Anti-Nuclear Contact Book*, published in 1977, listed 1,500 addresses of Citizens' Initiative and antinuclear groups.[30] That autumn, a demonstration in Bonn, the federal capital, attracted 150,000 protestors. The popular singer and songwriter Walter Mossman helped the antinuclear movement to its first success by opposing the Whyl nuclear project in the upper Rhine valley. It was never built. The protests were about more than nuclear energy. "The citizens' initiatives against nuclear power are part of an anti-war movement," Mossman observed in 1977.[31]

What was it about Germans and nuclear power? The sheer scale of their reaction was almost unique among Western democracies. Benny Peiser, who attended the conference that led to the founding of the Green Party and is now a leading climate skeptic, says that antinuclear sentiment had been seeded by the Far Right in the 1950s and 1960s. There was the philosopher and Nazi Party member Martin Heidegger and his "extreme technophobia." But there was something deeper, something seldom mentioned, Peiser says. Unconsciously, Germans did not understand why the Americans had not dropped an atom bomb on Berlin rather than Hiroshima as punishment for Germany's terrible crimes. It was a mixture of fear and guilt. Images of Hiroshima were often used by the antinuclear movement. Auschwitz and Hiroshima were deemed moral equivalents. In part, the fear came from the alleged hubris of playing around with plutonium and uranium nuclei to release vast amounts of energy—if a power station blew up, half the country would be gone. In part, it stemmed from the real threat of war and Warsaw Pact forces pouring through the Fulda Gap. As Peiser puts it, the West Germany of the late 1970s and early 1980s was enveloped by *Weltuntergangstimmung*—an apocalyptic mood, a presentiment of the end of times.[32]

The wave of Far Left terror had left the Sixty-Eighters marooned. There followed what Fischer called his "years of disillusionment."[33] He took up drums, swallowed hallucinogenic mushrooms, and tried to pay his bills by working as a taxi driver and in Cohn-Bendit's radical bookshop in Frankfurt. At first, the implications of the antinuclear protests—made up of reactionary peasants and petit bourgeois elements—passed them. It was Cohn-Bendit who saw the opportunity—and decided to join and lead the antinuclear movement. Once Fischer and the others saw it might get them elected and have proper jobs, they leaped on the bandwagon too.

The Green Party was formed in the first month of the new decade. As well as Cohn-Bendit and his comrades, the new party attracted the ecological old guard—ex-Nazis, neo-Nazis, and other Far Right nationalists. One of these was August Haussleiter and his AUD (Commonwealth of Independent German Action) party. Founded in 1965, the AUD developed elements of Nazi ideology, notably its environmentalism, anticonsumerism, and promise of a Third Way between communism and capitalism. Thanks to the antinuclear protests, it made some electoral headway in Bavaria. After the formation of the Greens, Haussleiter was elected a Green Party spokesman and the AUD dissolved. What also united the old and new Greens was neutralism. Among those attracted to the antinuclear cause, Hülsberg sees a rapid evolution from hostility to nuclear power to broader acceptance of environmentalism. "From the discussion of nuclear power it was a short step to considering the fragility of the whole eco-system and the possibilities of alternative sources of energy," he wrote in 1988.[34]

NATO's decision to deploy Pershing missiles was the force multiplier for the antinuclear movement. In 1976, the Soviet Union had begun deployment of SS-20 intermediate nuclear missiles. Deployment of SS-20s was a knife threatening to sever the Atlantic Alliance, as they exposed Western Europe to the threat of being decoupled from the American nuclear umbrella. That year, Helmut Schmidt warned that strategic arms negotiations between the United States and the USSR that ignored the balance of nuclear weaponry within Europe would impair the security of Western Europe. In December 1979, the Atlantic Council agreed to deploy Pershing II and cruise missiles to counterbalance the SS-20s. At the same time, the alliance would negotiate mutual limits on both sides' arsenals of such weapons to render deployment unnecessary.

The Atlantic Alliance's double-track decision unleashed massive protests and an intense Soviet propaganda offensive to generate sufficient public opposition to block deployment. Its fulcrum was West Germany: America's most important NATO ally, the battleground of any actual military conflict, and site of a preexisting cultural hostility to nuclear power, let alone nuclear weaponry. West Germany was also heavily penetrated by East German agents. From Schmidt down, West Germany's political leadership understood what was at stake. As Konrad Seitz, the head of planning at the West German foreign ministry, explained in 1982:

> Today the primary danger in Europe is not aggression and open warfare. We risk rather to see permanently modified, to the advantage of the Soviet Union, the force balances in Europe and the world. At the end of this process, the European democracies would see themselves constrained to self-neutralization. The Soviet Union would have won political control over Western Europe without having had to fire a shot.[35]

The Soviet leadership saw West Germany as the strategic pivot: if successful, the West would lose the Cold War. The two most prominent interviews by Soviet presidents on intermediate nuclear missiles (Brezhnev in November 1981 and Yuri Andropov in April 1983) were given to *Der Spiegel*. Although protests against new nuclear power stations and the maneuvering that led to formation of the Greens predated the advent of the Peace Movement, the two became enmeshed. By 1981, according to one survey, 70 percent of active supporters of the Peace

Movement were also supporters of the Greens.[36] The Soviets financed the Peace Movement, and Communist Party members were among its most active organizers. According to Peiser, the Greens bought into the Peace Movement and made it a mass movement.[37] In turn, the Peace Movement helped turn German politics green.

The first anti-NATO demonstration in 1981 attracted 250,000 people. A year later, 300,000 people demonstrated against the NATO summit in Bonn. The *New York Times* reported how the DKP had "dominated and manipulated" the planning of the demonstration. "The Communists dominated the meeting completely," a Green activist complained. "It took place under seemingly democratic rules, but that was a joke." Delegates approved a resolution describing the goal of the NATO summit as "support of the Reagan Administration's attempt to achieve worldwide hegemony," whereas the goal of the Greens, its leader Petra Kelly said, was a "non-aligned movement" and a Europe without nuclear weapons.[38] In the autumn of 1983, more than a million people took to the streets, more than half a million in Bonn alone. "What we are now experiencing is nothing less than the struggle for the life or death of society," the charismatic Kelly declared. [39]

"Without the NATO missile deployment, the Greens might have remained an interesting, offbeat group on the fringe of German politics," the American journalist and environmentalist Mark Hertgaard wrote.[40] The Greens called for a step-by-step detachment of West Germany from NATO and proclaimed their equidistance between the United States and the USSR—"Not to the East, not to the West, but loyal to ourselves." West Germany would defend itself with "non-violent social defense," just as the Norwegians had resisted the Quisling regime in 1940 and the Czechs had done after the Prague Spring in 1968. It was infantile, but it drained support from the SPD. To Brandt, the Greens were the SPD's lost children, and he gave the Greens reason to believe he was on their side. "I don't want that stuff here," he said of the NATO missiles.[41]

The irruption of the Greens into German politics had immediate and far-reaching consequences. The left wing of the SPD became more powerful. Schmidt's authority was undermined, as was the reliability of Social Democratic ministers and MPs. Schmidt fell in October 1982 when the SPD's junior coalition partner, the centrist Free Democrats (FDP), switched sides to Helmut Kohl's Christian Democrats. Electorally, the Greens pulled the SPD leftward. As historian Timothy Garton Ash notes, "Those active in the peace and ecological movements could threaten, explicitly or implicitly, to abstain or vote for the Greens."[42] In the 1983 federal elections, the Greens won 5.6 percent of the vote and Kelly became one of 27 Green MPs.

Kelly had spent the 1960s in America (her stepfather was a U.S. serviceman). She worked on Bobby Kennedy's, then Hubert Humphrey's, presidential campaigns and studied at American University in Washington, D.C. Her political heroes were not Marx and Lenin but Henry David Thoreau, Martin Luther King Jr., and anti-Vietnam war campaigners like Daniel Ellsberg and the Berrigan brothers, Daniel and Philip. Despite—or, perhaps, because of—that exposure, a month before her election, she held a war crimes trial in her Nuremberg constituency that put the United States in the dock. The *New York Times* reported that the mock trial

> drew strong parallels between Auschwitz and Hiroshima, belabored the United States for its purportedly aggressive nuclear weapons posture and, in the main, had little critical to say about the Soviet Union. The event, in the view of some, was shot through with the Greens' left-wing nationalism that casts West Germany as victimized by the United States and hobbled by its NATO connection.[43]

It was different on the other side of the Berlin Wall. Bahro recounts what happened when Kelly and her companion, Gert Bastian, an ex-NATO general who had quit to head the Peace Movement's Generals for Peace and Disarmament, and other Green activists were arrested in East Berlin's Alexanderplatz. As soon as the East Berlin authorities realized what was happening, they began to apologize.

> The next day, a letter arrived from [Erich] Honecker, personally addressed to our people, thanking them for the visit and saying that he was willing to make the GDR a nuclear-free zone. . . . Honecker, then, really does seem interested in the development of the Greens here.[44]

Honecker was very interested in the Greens and knew a lot about them. Bastian was a Stasi agent.

The Greens further increased their votes in the 1987 federal elections, nearly overtaking the FDP. But the first elections of the newly reunified Germany in December 1990 were a disaster, and the Greens lost all their seats. Two years later, Bastian shot Kelly and then himself. His last phone call had been in response to an urgent inquiry from the Greens to give the party permission to inspect his Stasi file.[45]

The Soviet bloc's penetration and manipulation of the Peace Movement is scarcely surprising as the two had the same objective: to prevent deployment of Pershing and cruise missiles. Any criticism of the Soviet Union was blocked, and detailed inner knowledge of the Peace Movement enabled Soviet propagandists to tune their messaging to sympathetic audiences in the West. Writing in *Commentary* in 1982, the Soviet dissident Vladimir Bukovsky remarked that he had not been surprised when "a mighty peace movement came into being." After 34 years of living in the Soviet Union, "I have some sense of its government's bag of tricks, pranks and stunts."[46] At the start of the Cold War, it had set up the World Peace Council, and in 1981 it established the International Peace Coordination and Cooperation network in Copenhagen (not to be confused with the climate change IPCC). Arne Petersen, an IPCC coordinator and a leader of the Danish peace movement, was caught funneling money from the Soviet embassy to the Peace Movement and arrested in a fur coat paid for by the KGB.

Three months after becoming chancellor, Kohl explained why West Germans would stand firm.

> The great majority of our people understand the issue and share my conviction that we must not be open to [Soviet] blackmail and pressure. . . . For us peace and freedom for the Federal Republic is the dominant issue. Everything else takes second place.[47]

History proved Kohl right. But ending the Cold War on the West's terms came at a cost. In a prescient paper written in 1987, Professor David S. Yost of the U.S. Naval Postgraduate School commented that the "affair has served as a vital socialization experience for many members of the successor generations in West Germany. . . . The INF [Intermediate-Range Nuclear Forces] affair may be expected to leave important and enduring political aftereffects."[48]

In December 1985, a tieless Joschka Fischer in running shoes was sworn in as minister of energy and the environment in the state of Hesse in Germany's first Red–Green coalition. In Hesse, a state that includes Frankfurt, Fischer began the phase-out of nuclear power. By 1997 there were nine Red–Green coalitions at state level, and the following year saw the first federal one. Fischer became vice chancellor and foreign minister; Otto Schily, who had left the Greens and joined the SPD, was minister of the interior; and an ex-communist Green, Jürgen Trittin, was put in charge of the environment and nuclear safety ministry. Germany—and Europe—would never be the same again. The Greening of Europe was the price the West paid for winning the Cold War.

10

Scientists for Peace

The exact consequences of an exchange of nuclear weapons would depend on many factors. Among the most important would be the weather; winter would be a particularly cruel time for those who managed to survive.

Palme Commission on Disarmament and Security, 1982[1]

The central point of the new findings is that the long-term consequences of a nuclear war could constitute a global climatic catastrophe.

Carl Sagan, 1983[2]

Can it be supposed that science cannot make a difference in the one matter that transcends all the others? This is not a conclusion that scientists will swallow.

William D. Carey, executive officer of the AAAS and publisher of Science[3]

In the brief period between the end of World War II and the onset of the Cold War, scientists entered the public arena: Only a world government could prevent an atomic war. They would reenter the nuclear arms debate during the Cold War's final decade with a similar naiveté and penchant for catastrophism.

All other problems paled into insignificance, Nobel-winning chemist Harold Urey declared two months after the first atom bomb.[4] In 1946, major American periodicals carried over 300 articles on the atom bomb.[5] Scientists were treated like a secular priesthood, endowed with foreknowledge of the end of things yet motivated by the highest ideal to save humanity from itself. The University of Chicago's

Edward Shils, friends with leading figures in the scientists' movement, commented that scientists' activism was rooted in their conviction that "they alone possessed an awful knowledge which, for the common good, they must share with their fellow countrymen, and, above all, with their political leaders."[6] As Urey put it,

> It is as if a bacteriologist had discovered a dread disease which might lead to a disastrous epidemic. He would not be a "politician" if he asked that the city health commission take measures to deal with a plague. He would merely be demonstrating common decency and social awareness.[7]

The *Bulletin of the Atomic Scientists* was launched in December 1945 with the hands of the clock close to an apocalyptic midnight. Scientists had continual access to columns of the *New Republic* and *Saturday Review*, and the services of a ghost writer and publicist. The Emergency Committee of Atomic Scientists raised $100,000 in the summer of 1946. "Nobel Prize winners have made themselves nearly as accessible as politicians," said one journalist.[8] "What a physical scientist says on almost any subject is thought more important that what anybody else says," observed the anthropologist Robert Redfield.[9]

As historian Paul Boyer wrote in his account of the cultural and intellectual impact of the atom bomb, scientists' calls for the immediate establishment of world government "revealed a profoundly ahistorical consciousness."[10] Such was scientists' immense prestige at the time, their efforts resulted in the Acheson-Lilienthal report, written in large part by Robert Oppenheimer. The Acheson-Lilienthal plan called for the internationalization of atomic power, with countries being given licenses to pursue peaceful nuclear research. It relied on Soviet–American cooperation and implied that the United States would destroy its nuclear arsenal. A modified version was rejected by the Soviet Union in 1946. Boyer describes the initiative as deluded, naïve neo-Wilsonianism that foundered on the rock of national sovereignty.

The plan's demise marked the high-water mark of scientists' influence in the immediate postwar period. The longer-lasting impact of the atomic bomb was in putting a question mark over science and technological progress. In a 1955 lecture, Isidor "I. I." Rabi remarked that the pursuit of science, which was once held in great esteem, was now equated with the "destruction of life and the degradation of the human spirit." Many had reached the glib conclusion that "science and the intellect are therefore false guides" and were turning elsewhere for "hope and salvation."[11]

There were subsequent spikes in the environmental consequences of nuclear weapons. The H-bomb tests on Bikini Atoll showed the impact of radiation on coral reefs. Radioactive rain fell in Chicago in 1955, and in 1959 strontium-90 was found in cows' milk. The 1963 Test Ban Treaty led to a reduction in antinuclear sentiment before reviving in the early 1980s. On June 12, 1982, nearly three-quarters of a million people took part in a nuclear freeze protest in New York's Central Park.

The same month, the Swedish Academy of Sciences published a paper, "The Atmosphere After a Nuclear War: Twilight at Noon," by two atmospheric scientists, Paul Crutzen and John Birks, in its *AMBIO* journal. Smoke from uncontrolable fires caused by a nuclear exchange would send 200 million to 400 million metric tons of particulate matter into the atmosphere. A large fraction of the solar radiation would be screened out for many weeks, "strongly reducing or even eliminating" the possibility of growing agricultural crops over much of the Northern Hemisphere.[12] Rain would be highly acidic. Darkness would kill off most of the plankton in more than half the oceans of the northern hemisphere. The effect would be similar to that from the impact of the asteroid that triggered the mass extinctions at the end of the Cretaceous period around 65 million years ago, which wiped out the dinosaurs, the paper claimed.

The authors didn't venture a firm view on the possible long-term climatic effects of a nuclear war. The eruption of Krakatoa in 1883 had injected similar quantities of aerosols into the atmosphere to what they reckoned would be caused by nuclear war, and global mean temperatures had been affected for only a few years. Nonetheless, the message of their paper was chilling. Because of fires, "it is likely that agricultural production in the Northern Hemisphere would be almost totally eliminated, so that no food would be available for the survivors of the initial effects of the war."[13] The atmospheric effects of nuclear war were complex and difficult to model, the authors concluded. "It is hoped, however, that this study will provide an introduction to a more thorough analysis of this important problem."[14]

It certainly did. That month, executives from the Rockefeller Family Fund met with the president of the national Audubon Society, whose wife happened to be editor of *AMBIO*, to discuss it. They contacted astronomer Carl Sagan, who was at the height of his fame narrating a television voyage through the cosmos. Sagan reported that a small group of scientists had been working on the climatic effects of nuclear war. In view of the *AMBIO* paper, Sagan's study was being broadened to include soot and smoke.[15] Two papers were then prepared. The first was on the physics, the second on the impacts on the biosphere. Before they were published, a two-day conference in Washington, D.C., would be held at the end of October 1983. The purpose was political. Scientists and environmental NGOs were putting tanks on the lawn of the White House.

In March, President Reagan had given a major address on nuclear weapons and national security. He reviewed the Soviet nuclear weapons buildup, explained the rationale for deterrence, and spoke of the nuclear imbalance in Europe (there were 1,300 Soviet warheads on SS-20s to none for NATO). He took on the arguments for a nuclear freeze. It would reward the Soviets for their massive military buildup and remove the incentive for them to negotiate arms reductions. Reagan then made the case that the world would be safer if it could rise above a peace based on the threat of mutual destruction. To this end, the President announced a long-term research-and-development program to render strategic nuclear missiles "impotent and obsolete." He acknowledged that what became known as the Strategic Defense Initiative (SDI) was a formidable task that might not be accomplished before the end of the century but could pave the way for arms control talks to eliminate the weapons themselves. "Our only purpose," the President concluded, "one all people share—is to search for ways to reduce the danger of nuclear war."[16]

SDI galvanized Sagan. The day before the Washington conference, he had a lengthy article on his nuclear winter hypothesis in *Parade,* a mass-circulation Sunday magazine ("It was in *Parade* that astrologer Jeane Dixon predicted the assassination of JFK").[17] At the end of the article, readers were urged to write to Presidents Reagan and Andropov to support arms reductions or a nuclear freeze. Andropov would surely have been pleased: The only conceivable practical effect of the nuclear winter was to undercut the West's ability to negotiate mutual arms reductions.

The conference itself was supported by 31 groups that today read as a roll call of organizations at the forefront of the climate wars—Common Cause, Environmental Defense Fund, Friends of the Earth, International Union for the Conservation of Nature, Audubon Society, Natural Resources Defense Council, Planned Parenthood, Union of Concerned Scientists (formed in 1968 to campaign against the Vietnam War and the development of an anti-ballistic missile system), World Resources Institute, and Zero Population Growth. A satellite link from the Kremlin was cofunded by the Tides Foundation, a cash bundler of progressive causes. In addition to media-savvy scientists such as Stephen J. Gould, Carl Sagan, and Jared Diamond, the conference's advisory board included the two *AMBIO* authors; Olof Palme's long-time scientific adviser, Bert Bolin; and three representatives from the Soviet Union (including the Soviet minister of health).

As well as scientific personnel, NGO infrastructure, and the UN (the United Nations Environment Programme was in on the act), nuclear winter shared other features with global warming—hyperbole, overstatement, reliance on unverifiable model projections masquerading as empirical evidence, and the endemic failing on the part of scientists to state what they don't know and couldn't know. "The scientific discoveries described in this book may turn out, in a world lucky enough to continue its history, to have been the most important research findings in the long history of science," Lewis Thomas, a cancer specialist and Sloan Kettering president, wrote in the foreword to the book of the conference proceedings.[18] It was an absurd statement. Nothing had been "discovered." Nuclear winter was a conjecture designed to generate acres of media attention—something it did very well.

The Washington conference was heavily trailed in the *New York Times*. "The lack of sunlight could cause a 'harsh nuclear winter' with temperatures dropping as much as 25 degrees centigrade in inland areas, the report said, adding that many areas could be subject to continuous snowfall, even in summer."[19] Sagan and Paul Ehrlich appeared on ABC *Nightly News*. *Time* magazine covered the conference papers under the heading "A cold, dark apocalypse." Toward the end of the session with the Moscow hookup, Thomas Malone, who had the distinction of being simultaneously president of the American Meteorological Society and the American Geophysical Union and was now foreign secretary of the National Academy of Sciences, remarked that the exchange of views between Washington and Moscow "may turn out in years ahead to be viewed—correctly—as the turning point in the affairs of humankind and will elevate the level of consciousness among policy-makers."[20] As a prediction, it was unmitigated drivel.

The conference wasn't about objective truth-seeking. Consciousness raising was Sagan's goal. "For me, the new results [*sic*] on climatic catastrophe raise the stakes of nuclear war enormously," Sagan wrote in *Foreign Affairs*, describing something that hadn't actually happened. "There is a real danger of the extinction of humanity."[21] The formal paper, known as TTAPS after the surnames of Sagan and his coauthors, was published in *Science* in December. Its first sentence cited the *AMBIO* paper, and its third picked up its suggestion of massive fires generating smoke that would affect the climate. These were then modeled, and TTAPS came to a "tentative" conclusion of a "harsh 'nuclear winter.'" As with acid rain and later with global warming, such tentativeness was short-lived. Although the estimates were uncertain, "the magnitudes of the first-order effects are so large, and the implications so serious, that we hope the scientific issues raised here will be vigorously and critically examined."[22]

The process of discarding uncertainties continued in the second paper on the biological effects, published in the same number of *Science* and written by Sagan, Ehrlich, and 18 biologists, including a future president of the American Association for the Advancement of Science (Peter Raven), a future president of the Royal Society and chief scientific adviser to the British government (Robert May), a founding trustee of the Natural Resources Defense Council (George Woodwell), and Stephen J. Gould. Their paper described the predictions of climatic changes as "quite robust."[23] Despite being restricted to the output of a one-dimensional model, the 20 scientists opined that it was possible that darkened skies and low temperatures would spread over the entire planet. "It seems unlikely, however, that even in these circumstances *Homo sapiens* would be forced to extinction immediately. Whether any people would be able to persist for long . . . is open to question."[24] Eventually there might be no human survivors in the northern hemisphere. The paper ended on a somber note, in bogus scientific prose crafted to cause alarm.

> Global environmental changes sufficient to cause the extinction of a major fraction of the plant and animal species on Earth are

> likely. In that event, the possibility of the extinction of *Homo sapiens* cannot be excluded.[25]

Such was the importance accorded the two nuclear winter papers that the publisher of *Science*, William D. Carey, wrote an editorial. Science was being used for political ends and national security purposes, Carey editorialized. When scientists realized what was happening, it produced "a level of discomfort," which then expressed itself "within the strictures of science's methodologies, in concerted displays of scientific responsibility." It had been a "good thing" for scientific integrity and a sign of courage that some 40 scientists of high standing had spoken on the long-term effects of nuclear war.[26] In reality, Carey had it entirely the wrong way round. Nuclear winter was policy-driven science directed against the Reagan Administration and the strategic interests of the West. Seventeen of Sagan's 36-page article in *Foreign Affairs* was devoted to nuclear policy and doctrine. "What is urgently required," Sagan concluded, "is a coherent, mutually agreed upon, long-term policy for dramatic reductions in nuclear armaments."[27] That goal had been outlined by Reagan in his March address. The difference was that Reagan knew what he was doing and Sagan—unless he was a Soviet stooge—did not.

A devastating riposte to Sagan by physicist Russell Seitz appeared in the next issue of *Foreign Affairs*.

> The mathematical model employed by Dr. Sagan and his colleagues gives us an internally coherent account of how black particles placed in the stratosphere *could* block sunlight. However the soot is placed there not by the model but by the modelers. The stratospheric soot in the TTAPS simulation is not up there as an observed consequence of nuclear explosions but because the authors told a programmer to put it there.[28]

Seitz kept on the offensive against the nuclear winter through the 1980s, using it as a cause célèbre of the politicization of science. In *Merchants of Doubt*, Naomi Oreskes and Erik Conway portray Seitz's critique as the start of a right-wing turn against science subsequently taken up by Rush Limbaugh in the 1990s and Michael Crichton in the 2000s: "The attack on nuclear winter was a dress rehearsal for bigger fights yet to come."[29] There is one not-insignificant wrinkle in this interpretation. When it came to nuclear winter, the biggest "merchant of doubt" was none other than the dedicatee of the synthesis report of the IPCC's Fifth Assessment Report—the much-lauded climate scientist Stephen Schneider.*

In a 1986 *Foreign Affairs* article with fellow climate scientist Starley Thompson, Schneider wrote that, on scientific grounds, "the global apocalyptic conclusions of the initial nuclear winter hypothesis can now be relegated to a vanishingly low level of probability."[30] The TTAPS model had no geography, no winds, no seasons, instantaneous spread of smoke to the hemispheric scale, and no feedback of atmospheric circulation changes on the rate of smoke washout by rainfall. TTAPS had postulated a megatonnage climate tipping point to argue that the world's total nuclear arsenal should be kept below it. Schneider and Thompson said the concept was not "scientifically persuasive."[31] Several of the strategic implications of nuclear winter, as originally conceived, "are no longer justified."[32]

A similar conclusion was reached by Fred Singer in a coruscating 1988 critique of a quantified version of TTAPS in a paper produced by the National Academy of Sciences in 1985. Baseline values for the amount of smoke had been fine-tuned to lead to surface cooling: Much more, and saturation would occur, delaying cooling; a little less, and at low altitude, then surface warming would result. "The assumptions underlying the TTAPS study virtually guarantee the occurrence of a 'nuclear winter,'" Singer wrote. Any cooling effect was likely to be short-lived because of the sweep-out of smoke clouds from larger dust particles, rain droplets, and snow. Together with the injection of more greenhouse gases into the atmosphere, these effects might even induce slight warming—a mild "nuclear summer," Singer dubbed it—something that would not have fitted with the alarmism of the politically desired narrative.[33]

At the end of their paper, Schneider and Thompson cut to the chase by placing the whole issue in its proper context. In doing so, they displayed a maturity wholly lacking among the vast majority of their scientific peers. The problem of avoiding nuclear war was "not amenable

* Oreskes and Conway mention a *Nature* editorial by Maddox in the main text of *Merchants of Doubt* and reference Schneider and Thompson's *Nature* paper but do not mention that their analysis demolishes the scientific claims of the nuclear winter hypothesis.

to scientific solution," they argued.[34] "Relative to official US reticence, the Soviet Union has, to many Western observers, appeared steadfast in its ostensible acceptance of the absolute horrors implied by the earliest nuclear winter scenario."[35] The Soviet position might have been adopted for its propaganda value "presumably in the hope that US and West European public opinion could be used to pressure the United States into arms control negotiating positions more favorable to the Soviets."[36] Exactly.

The funeral rites for the nuclear winter were performed in a carefully worded 1988 editorial in *Nature* by John Maddox in a way that implied that, as science, it was not completely dead. Schneider and Thompson's paper was not "the end of nuclear winter but a proof (if one were needed) that nuclear winter has come of age," Maddox wrote.[37] The original *AMBIO* paper "stood impeccably on solid ground by publishing their calculation of what might happen if there were a lot of smoke in the atmosphere." But the authors had not properly understood the effects of their "ratiocinations" on the world at large. All this on the strength of a one-dimensional model? "Many one-dimensional models are remarkably predictive," Maddox answered. "For the time being, at least, the issue of nuclear winter has also become, in a sense, irrelevant."[38]

It had been rendered irrelevant by Ronald Reagan in what Henry Kissinger calls "an astonishing performance."[39] The Kremlin had made opposition to deployment of cruise and Pershings the linchpin of Soviet foreign policy. Its dismissal of the "zero-zero" option (under which NATO would dismantle cruise and Pershings if Moscow did the same with its SS-20s) had, according to Kissinger, turned out to be a "stunning victory" for Reagan and Helmut Kohl.[40] The leaders of the West prevailed in the face of the most determined and sophisticated propaganda campaign since the fall of Nazi Germany. On December 8, 1987, Mikhail Gorbachev went to the White House to sign the Intermediate-Range Nuclear Forces (INF) Treaty to eliminate shorter and intermediate nuclear weapons, opening the way to the peaceful ending of the Cold War.

NATO leaders knew that failure to counter SS-20s would, over time, lead to the Soviet Union winning the Cold War. The Soviet leadership also knew what was at stake. By propagating the science of nuclear winter, American scientists and their foundation and NGO allies acted to promote the strategic interests of the Soviet Union in aiming to defeat the West in the Cold War. If they had looked, there were strong clues that the *AMBIO* paper was a Soviet–Swedish coproduction. The foreword to the American edition was written by Alva Myrdal, and the British edition was published by Robert Maxwell's Pergamon Press, publisher of classics such as *Leonid I. Brezhnev: Pages from His Life,* by Leonid I. Brezhnev (1982).

In 2000, a senior agent in the KGB's successor agency defected. According to the FBI, Sergei Tretyakov "literally held the keys to a Russian intelligence gold mine."[41] At the KGB's Red Banner Institute, Tretyakov read dozens of case studies wherein the KGB had used propaganda and disinformation. There was one the KGB was most proud of: "It created the myth of nuclear winter."[42] After NATO's double-track decision, Andropov ordered the Soviet Academy of Science to produce a doomsday report to incite more demonstrations in West Germany. Soviet scientists had been researching dust storms and found an "anti-hot house effect" due to dust particles shutting out the sun's rays, a story which was duly picked up by the BBC. Under the direction of Yuri Izrael (a future vice chair of the IPCC), they applied the "anti-hot house effect" to the dirt and debris thrown up by the use of nuclear missiles in Germany during a Soviet attack to show that temperatures across Europe would plunge. "I was told the Soviet scientists knew this theory was completely ridiculous," Tretyakov said. "There were no legitimate scientific facts to support it. But it was exactly what Andropov needed to cause terror in West."[43]

Instead of publishing the fake findings in a Soviet scientific journal, Andropov decided to use the KGB's tried and tested "covert active measures." According to Tretyakov, KGB officers disseminated the study's conclusions to their contacts in the peace and antinuclear movements and in environmental NGOs. One of the publications they targeted was *AMBIO*, which then approached Crutzen.[44] Propagators of the nuclear winter thus acted as dupes in a disinformation exercise scripted by the KGB calibrated for maximum media impact, just as Andropov had intended. Even before the TTAPS paper had been peer reviewed, an $80,000 retainer had been paid to a Washington, D.C., public-relations firm.[45] Useful idiots indeed.

The nuclear winter scare was an objective test of scientific integrity and judgment. The number of climate scientists who criticized the nuclear winter hypothesis and subsequently endorsed global warming can be counted on the fingers of one hand and were vastly outnumbered by those who failed it.* Given the poor judgment of scientists of high standing, their credulity in promoting flimsy science, and the role of NGOs in propagating it and progressive foundations in funding it, prudence would, at the very least, suggest the greatest wariness when they came around with the next one. After all, why trust people who'd got the outcome of the Cold War so grievously wrong?

It was not to be.

On February 24, 1994, Ted Koppel on ABC's *Nightline* gave airtime to Al Gore and his allegations about who was funding climate skeptics. Koppel then went on to air footage from a previous *Nightline* about the climatic effects of the 1991 Kuwaiti oil fires. Fred Singer had predicted the smoke would dissipate quickly; Carl Sagan predicted massive environmental damage. "The record shows that in this instance Dr. Sagan was wrong and Dr. Singer was right," Koppel intoned, before going on to make an observation of continuing relevance.

> There is some irony in the fact that Vice President Gore, one of the most scientifically literate men to sit in the White House in this century, that he is resorting to political means to achieve what should ultimately be resolved on a purely scientific basis.[46]

Concluding his 1988 *Nature* editorial, Maddox alluded to the role of trust in evaluating global warming and the greenhouse effect. "Even when meteorologists are able to put their hands on their hearts and say that they have detected signs of global warming, it will fall to the model-builders to say what the consequences will be."[47]

The beauty of global warming for environmentalists is that the matter is not resolvable in the way Koppel wanted. It does not yield a clear-cut evidential test, as had Sagan's prediction of environmental catastrophe from burning oil wells that had enabled people to decide whether alarmists like Sagan or more empirically minded scientists like Singer were right, or, as had been the case with acid rain, whether

* Apart from Schneider, MIT's Kerry Emmanuel is the other prominent climate scientist critical of the nuclear winter hypothesis.

forests were rapidly dying, as the scientific consensus had falsely asserted. Instead, the science of global warming would be decided on the scales of a preponderance of scientific opinion, a process inherently prone to bias, in which national scientific academies and official bodies, especially the IPCC, would have a crucial role.

Nuclear winter shows that scientists acting as a group in a political context are not worthy of trust. It was not the first occasion when scientists accepted bias, disregarded confounding evidence and alternative hypotheses, and sacrificed scientific discipline for the sake of a political agenda—and it would not be the last. The scientific consensus on acid rain, as promulgated by the national academies, turned out to be wrong. In 1972, eminent British scientists had put their name to a document stating that the termination of modern industrial civilization within a lifetime was "inevitable."[48] Nuclear winter had been part of a KGB-inspired disinformation campaign. It amounts to a catalogue of embarrassing failure that, if made by any other profession, would have disqualified its practitioners from being taken seriously in political matters. Like their predecessors after Hiroshima, who had advocated world government, the proponents of the nuclear winter scare ignored the realities of the Cold War. In turn, their prescriptions of how the planet should be managed were ignored while the Cold War lasted.

Their historical incompatibility was recognized in a 1982 report by a UN panel of experts on disarmament and development chaired by Inga Thorsson, an undersecretary at the Swedish ministry of foreign affairs. The arms race could be pursued or the world could move "consciously and with deliberate speed toward a more stable and balanced social and economic development," the panel said. "It cannot do both." This was not just a matter of the resources diverted from development to fund the arms race but also of the "vital dimension" of attitudes and perceptions.[49] If the arms race continued and the principles of the UN Charter were not observed, the adoption of what the panel called the "co-operative management of interdependence" was "quite improbable."[50]

By trying to undermine Western solidarity, Sweden had done nothing positive to help bring about the end of the Cold War. It would have an outsized role in shaping the post–Cold War era in which scientists would have more influence and power than they had ever known.

11

Sweden Warms the World

As a result of the increasing concentrations of greenhouse gases, it is now believed that in the first half of the next century a rise of global mean temperature could occur which is greater than any in man's history.

UNEP/WMO/ICSU Climate Conference, Villach, October 1985[1]

There is no way to prove that any of this will happen until it actually occurs. The key question is: How much certainty should governments require before agreeing to take action? If they wait until significant climate change is demonstrated, it may be too late for any countermeasures to be effective against the inertia by then stored in this massive global system.

Brundtland Report, 1987[2]

In November 1974, Sweden's main conservative newspaper, the *Svenska Dagbladet*, asked the leaders of the Social Democrats and the Moderate Party about what would happen over the next 25 years to 2000. Both Olof Palme and the Moderate Party's Gösta Bohman were optimists—for completely different reasons.

Palme led off. People would be less materialistic and their lives less stressful. It meant first overcoming the "global crisis" by reducing the gap between rich and poor countries and by stopping the arms race. At home, there would be quite dramatic changes. "The more ruthless and brutal effects of capitalist society will be eliminated step-by-step," he predicted, criticizing consumerism for using up too much energy and resources. Although material advances wouldn't be as great as in the past, society would be more compassionate. So far, this was boilerplate Palme. Then he said something really interesting. What crisis is worrying you the most? "The risk of climate change from human activities."

> Prominent experts fear that if we continue on the same path as now, over the long term, the oceans might be destroyed, the Arctic ocean melt and other changes in the climate occur and drastically change living standards.

"For me, this issue is extremely important," Palme added. Solving it would require a new way of looking at the role of nation-states.

> We must overcome and arrive at a new epoch where we all realize that mutual inter-dependence is so important that we have to give up part of the national sovereignty. . . . We cannot solve these long term problems without democratic socialist ideas.[3]

The interview concluded with some political philosophizing. In some ways, Sweden was already the utopia it would be in 2000. A vision of utopia has the same function as a mirage in the desert, Palme explained. Without the mirage, you wouldn't get to the next oasis. What did Palme mean? Many political leaders are really boring except for the office they hold, Henry Kissinger once said. That wasn't true of Palme, whom Kissinger found interesting whether in or out of office.[4] Earlier in his time as prime minister, Palme said people had a right to be dissatisfied.

> I tell them that their enemy's reality, that they've got to fight it. Then we're all on the same side.[5]

Was global warming a political mirage for Palme to drive the camel train to the next oasis? Whatever his motive in raising it, Palme was the first serious politician to talk about climate changes as a serious threat, beating the runners-up by more than a decade.

Bohman's prediction was also remarkably prescient. By 2000, the demands of the peoples of Eastern Europe for freedom and democracy would lead to the collapse of communism. As a result, Europe would be reunited.[6] Neither man could have known that Palme's prediction depended on Bohman's being right: The ending of the superpower rivalry that had riven the post-1945 world was a precondition for global warming being able to define the post–Cold War era.

Back in the reality of Sweden in the 1970s, all was not well. There had been a peasants' revolt in the 1973 election, which saw a dip in support for the Social Democrats and a surge for the antinuclear farmers' Center Party. Sweden's "Gröna Vågon"—Green Wave—was a reaction against Social Democratic technocracy and its alignment with the interests of big industrialists. Although the election left government and opposition parties with the same number of seats, Palme pushed ahead with the Social Democrats' huge nuclear power program. It would give Sweden the highest per capita level of electricity from nuclear power anywhere in the world. In 2007, the average Swede would consume 19.6 kWh a day of nuclear power, just ahead of France (19.0 kWh per day) and far more than the United States (7.5 kWh per day) and Germany (4.4 kWh per day), the other country where nuclear power was a potent election-deciding issue.[7]

The politics of nuclear power played out against Sweden's deteriorating economy—the worst of any Western nation in the 1970s. Inflation rose, huge wage increases led to a big loss in competitiveness, and public-sector employment exploded. Sweden would end the decade plagued by strikes and lockouts and the highest taxes of any industrial nation, taxes needed to fund public spending that would mushroom to 64 percent of GDP.[8] For a prime minister to talk about global warming—global temperatures had been declining since the mid-1940s—as the issue that worried him most when the economy was falling apart might appear senseless. Palme was anything but a political naïf. He had no economic record to run on, which made it even more important for the Social Democrats to avoid being swamped by the Green Wave. The opposition parties were split on the nuclear issue, and the 1973 oil price shock strengthened the case for the nuclear power program to reduce Sweden's dependence on imported energy. This was a game Palme could play and win.

As on acid rain, Palme deployed Bert Bolin to front the science. In February 1975, Bolin produced *Energy and Climate*, described as a survey of current knowledge about how energy use could affect the Earth's climate.[9] The 54-page paper was planned as the first in a series by a government-controled think tank established by Palme. In 1975, a Swedish government bill on energy policy stated, "It is likely that climatic concerns will limit the burning of fossil fuels rather than the size of natural resources."[10] The party's program for the 1976 election stated that the threat to nature and the environment increased the urgency for economic planning. "Unfettered competition for natural resources will undermine the long-term possibility of giving everyone a good natural environment."[11]

The September 1976 election came as a shock. For the first time in 40 years, the Social Democrats were not in government, and Center Party leader Thorbjörn Fälldin became prime minister of an unstable coalition hopelessly split over nuclear power.

In the run to the 1976 election, the nuclear industry had helped Palme by playing up the idea that opposition to nuclear power would lead to Sweden having to switch to coal. Swedes had been taught that coal caused acid rain and that Scandinavia's lakes and forests were being destroyed. Sweden's hydropower potential had mostly been used up. To stoke alarm, the nuclear-power industry talked about converting oil-fired plants to coal and building new coal-fired power stations. Climate change was used in the anticoal campaign. The April 1975 issue of the nuclear consortium's *OKG-Aktuellt* news magazine reported that a delegation of technicians and researchers had presented Palme with a petition complaining about attacks on the industry. Unlike nuclear waste, which was highly concentrated and contained, fossil fuel combustion could not avoid depositing emissions into the atmosphere. The effects of carbon dioxide could be serious and long-term.[12]

The nuclear industry continued its campaign after the election. A January 1978 paper in *Nature* by J. H. Mercer of Ohio State University's Institute of Polar Studies caught the attention of *OKG-Aktuellt*. Under the headline "Catastrophic Climate Change on the Way?" the article reported that, on current trends, the amount of carbon dioxide in the atmosphere would double in around 50 years. This could lead to quick and catastrophic melting of the West Antarctic ice sheet and a five-meter rise in sea level.[13]

These messages were reinforced by the Social Democrats. The report of the party's Energy Policy Working Group, published in 1979, stated that sulfur emissions from coal caused acidification of lakes and soil, released toxic heavy metals, and put carbon dioxide into the atmosphere, which affected the Earth's climate. It was not practical to remove the carbon dioxide; such a technology would require more energy than produced by the power station.[14]

On March 28, 1979, one of the reactors at Pennsylvania's Three Mile Island nuclear generator suffered a partial meltdown. A week later, Palme held a press conference to announce that the Social Democrats would hold a referendum on the future of nuclear power after the election, which was due six months later. Palme's gambit worked, but its biggest beneficiary was not the Social Democrats, who gained a couple of seats, but the pronuclear Moderates, who won an extra 18, while the antinuclear Center Party lost 22 seats. Despite being overtaken by the Moderates, Fälldin returned at the head of another unstable coalition.

Held in March the following year, the nuclear power referendum was rigged by giving voters three options designed to maximize the chances of the Social Democrats' favored Option 2. Both the Moderate's Option 1 and Option 2 envisaged that nuclear power would be used until renewable alternatives became available (Sweden has lots of timber) and eventually phased out within 30 years. Option 2 clothed Option 1 in socialism. Nuclear power stations would become totally owned by the public sector (the state and municipalities already owned a majority), and the superprofits of hydropower generation would be subject to a punitive tax levy. Option 3, backed by the Center Party, proposed a moratorium on any expansion of nuclear power and all operational nuclear power stations be closed within 10 years.

In a big set-piece speech a month before the vote, Palme deployed the by-now tried-and-tested nuclear counterattack on fossil fuels. He compared the effects of nuclear power with oil and coal, especially the risk to human life and the environment. Fossil fuels had "direct and considerable effects" on the environment. First came acid rain. "Ten thousand lakes are already dead and another ten thousand are dying." Then there was global warming. "Carbon dioxide is changing the Earth's climate."[15] Palme's strategy prevailed. Option 2 gained the most support (39.1 percent), just ahead of Option 3 (38.7 percent), and far ahead of Option 1 (18.9 percent). Although the 30-year deadline passed in 2010, only two small reactors had been decommissioned and ten reactors were still operating.

Palme went on to win the September 1982 election, returning to power after six years in opposition. A year earlier, Palme's fellow Social Democrat, Gro Harlem Brundtland, ended her first stint as Norway's prime minister. In March 1982, UNEP head Mustafa Tolba had contacted Brundtland to sound her out about chairing a new UN commission on the environment modeled on the Brandt North-South commission and the Palme commission on disarmament and security, on which Brundtland had been a (pro-NATO) commissioner. The idea had been suggested to Tolba by one of Palme's boys, Ulf Svensson, "one of those young and enthusiastic Swedes who wanted a commission whereby one would really be able to review the range of problems that environmental difficulties imposed upon the world," Brundtland wrote in her memoirs.[16] At the end of the year, she was formally approached by UN Secretary-General Javier Perez de Cuellar to head the World Commission on Environment and Development. Her assignment was to establish and head the commission. "Sweden had promised to provide help with actual start-up, in the form of both expertise and financing."[17]

During the six years the Social Democrats were out of power, Swedish climate research lost its political patronage. Fälldin saw climate change as a ruse to promote the case for nuclear power. Bolin spent his time trying to internationalize global warming beyond Sweden and the parallel efforts being made in America by the National Academy of Sciences. The available information "did not yet stir public interest in the issue of possible human-induced climate change," Bolin wrote. Changing this required bringing together the political and scientific communities.[18] Bolin was an active force in the committee that set up the first joint conference of the World Meteorological Organization (WMO) and the UNEP in Vienna in 1979.

The first attempt at an international assessment of climate change took place under Bolin's chairmanship in Villach, Austria, in 1980. Bolin was disappointed at its lack of impact. What was needed, he thought, was something that went well beyond an analysis of the physical aspects. Would the WMO and UNEP support the undertaking? On the train from Villach, Bolin discussed the possibility of doing something more substantial with a group of conference participants and representatives of WMO and UNEP. "I expressed the view that an analysis that was wider in scope, greater in depth and more international was most desirable."[19]

Momentum picked up in June 1982 when Tolba visited Stockholm at the behest of Göte Svensson, a Social Democrat and former top bureaucrat in the Ministry of Land Planning.* UNEP stressed that the assessment should focus on the impact of climate change on global ecosystems. Acceptance of UNEP's requirement led to a formal agreement between UNEP and the University of Stockholm on an assessment project. "UNEP financial support was generous," and Bolin submitted a project plan and was given "great freedom" to carry out the work, which got under way toward the end of 1983.[20] The study resulted in a 500-page report published in November 1985. Chapters from it formed the basis of presentations at another meeting in Villach in October. According to John Zillman, director of the Australian Bureau of Meteorology and a lead participant in many of those early meetings, though not at Villach, it would be difficult to overstate the significance of the 1985 Villach conference.[21] Tolba gave "a very powerful message" about possible future disasters, and the conference concluded with a declaration that the first half of the twenty-first century could see a rise in average global temperature "greater than any in man's history."[22]

From a scientific viewpoint, what is most significant about the Villach claim is what's not there: any numbers. When you express in numbers what you are speaking about, the British mathematical physicist Lord Kelvin said, you know something about it; when you cannot, your knowledge is of a meager and unsatisfactory kind. The climate scientists at Villach had their numbers but chose not to quantify their headline-grabbing claim. This wasn't science; it was scare mongering. John Maunder of New Zealand, president of the WMO Commission for Climatology for eight years during the 1990s, was a participant at Villach. In 2012, he questioned the declaration's claim that changes in greenhouse gases were likely to be the most important cause of climate change during the twenty-first century.

> At that time, even though I was partly responsible for the writing of the above paragraph, I along with a few of my colleagues,

* Svensson is one of the top ten most common surnames in Sweden, but I have not been able to determine whether Göte and Ulf are relatives.

> had some misgivings about this phrase, and were somewhat surprised that within a year "human-induced global warming" caught the imagination of much of the world.[23]

Neither has Villach's "greater than any in man's history" claim weathered well. Based on the British Met Office's HadCRUT4 decadally smoothed series of average global temperature, in the last decade and a half of the twentieth century, the average global temperature rose at a rate of 0.213° centigrade per decade. Between 2000 and 2014, it rose at 0.112° centigrade per decade—less than one-third the 1985–2000 rate—and less than half the increase in the 15 years to the mid-twentieth-century peak in 1943.[24]

The conference recommended setting up of a six-person Advisory Group on Greenhouse Gases (AGGG), with WMO, UNEP, and the International Council for Science (ICSU) nominating two members apiece. It was chaired by Kenneth Hare, the Canadian meteorologist. Well versed in the environmental policy game, Hare had chaired the Canadian review panel on acid rain that in 1983 had opined that the facts on acid rain were "actually much clearer" than other environmental causes célèbres and asserted that the existence of a severe environmental problem caused by acid rain was "not in doubt."[25] If one scare wasn't enough, Hare was also chairing a Royal Society of Canada study into the nuclear winter. The by-now-inevitable Bolin was on the AGGG and joined by Gordon Goodman from the Stockholm-based Beijer Institute, which was an important funding channel for the effort—research money means power.* At the same time, Goodman was writing the energy chapter of the Brundtland report.[26] Bolin had also been asked by ICSU to chair a committee scoping a research program on the global dimension of chemical and biological processes. By the time it came to launch what became known as the International Geosphere Biosphere Program (IGBP) later in the year, Bolin was in an even better position to influence developments.

At 8 A.M. on April 26, 1986, abnormally high levels of radioactivity were recorded outside the Forsmark nuclear power station, 73 miles

* The Beijer Institute was established in 1977 by Kjell Beijer, owner of Kol & Koks AB (Coal and Coke Ltd.), a coal-, coke-, and oil-trading business founded in the 1920s. The focus of the Beijer Institute was on energy and human ecology. In 1989, its activities were merged with the state-run Stockholm Environment Institute.

north of Stockholm. The level of radioactivity was much higher outside the building than inside. It looked like there had been a major radioactive discharge. The Vattenfall head office, the power station's owner, was immediately alerted, as was the Swedish nuclear safety watchdog. Minutes later, a similar report was received from the Studsvik nuclear research center 45 miles south of Stockholm. Then a report came from the Oskarshamn nuclear plant. By 10 A.M., the whole nuclear chain in Sweden was on high alert. The prevailing easterly wind led people to speculate that there had been an incident somewhere in the Soviet Union, but they didn't know there was a nuclear power station at Chernobyl, which was controlled by the Soviet military. It took two days for the Soviets to admit there had been an accident. On May 1, Bolin was formally appointed the prime minister's scientific adviser, with an office in the Rosenbad building with the prime minister and his closest staff.

The IGBP was one of the first beneficiaries. "Being at the time scientific advisor to the Swedish prime minister," Bolin noted in his IPCC memoir, "I was able to secure financial support from the Swedish government to develop this international research program, and to suggest that the secretariat be located in Sweden."[27] The strands were coming together. From Stockholm, Bolin "informally channelled" the IGBP's preliminary results to the secretariat of the Brundtland Commission.[28] The flow was reciprocated when Jim MacNeill, secretary of the Brundtland Commission, provided advance briefing to the tenth World Meteorological Congress in May 1987 in Geneva.[29] Important as this was, such dialogue wasn't sufficient for Bolin: "An organ that provided an international meeting place for *scientists and politicians* to take responsibility for assessing the available knowledge concerning global climate change and its possible socio-economic implications was missing."[30]

In Zillman's 2007 account, the defining moment of the congress was an impassioned plea by the lead delegate from Botswana for a mechanism to provide her with an authoritative assessment on what was known about human-induced climate change. Intense corridor discussion followed, from which, we are asked to believe, the idea spontaneously emerged of an intergovernmental panel. It was exactly what Bolin wanted. Offline it was agreed that further informal consultations should be held with UNEP.[31] After the congress, the WMO and UNEP governing bodies passed resolutions to establish an ad hoc intergovernmental mechanism to carry out scientific assessments on the magnitude, timing, and potential impacts of climate change. Representatives of WMO and UNEP member states were invited to a meeting in Geneva in November 1988 to establish an Intergovernmental Panel on Climate Change. By then, preparations were already under way for a climate conference to be held in Toronto in June 1988 immediately following the G7 summit that Canada was hosting. The AGGG was invited to assist. "In this way I was given the opportunity to work with the organizing committee," Bolin recorded.[32]

The soon-to-be-formed IPCC couldn't be left to its own devices. With the institutional structure coming together, it became important to ensure that it would be steered in the right direction and produce the right answers. This was the purpose of two week-long workshops held in the autumn of 1987, the first in Villach at the end of September and the second in Bellagio on Lake Como in November. These meetings constitute important evidence as to the political nature and purpose of the embryonic IPCC. If the function of the IPCC were limited purely to collecting and evaluating scientific papers on climate change, why did its creators and cosponsors convene workshops on developing policy responses to climate change? Run-of-the-mill climate scientists couldn't be entrusted with the future of the project because they lacked an understanding of its deeper purpose. As the Bolsheviks knew, the revolution had to be guided by a revolutionary elite. In truth, the IPCC had a political agenda encoded in its DNA at its conception.

Held under the auspices of the Beijer Institute—other funders included the Swedish Energy Research Commission, UNEP, two Rockefeller foundations, and the German Marshall Fund—participation was much more restricted than the WMO congress in May: 48 at Villach and 24 at Bellagio, indicating that many—though not all—of those attending were core to the project.* In addition to Bolin and Goodman, at Bellagio, the scientists included two nuclear-winter warriors (Paul Crutzen and George Woodwell, both of the Woods Hole Re-

* The 50-page summary of the meetings produced by the Beijer Institute mentions the existence of a "Steering Committee of the project," although not all its members are identified. WMO/UNEP, *Developing Policies for Responding to Climatic Change* (April 1988), title page.

search Center) and the Environmental Defense Fund's Michael Oppenheimer, who was in the process of migrating from acid rain to climate change. On the policy side, there were representatives from the EPA, the European Commission, and Environment Canada; the Brundtland Commission's MacNeill; and Måns Lönnroth from the Swedish Cabinet Office, who had written the foreword to Bolin's first report on global warming 12 years earlier. They were joined by representatives of the in utero IPCC's sponsoring organizations, UNEP and the WMO.

As with the earlier Villach meeting, the Bellagio group was prone to overprediction. If the 1987 Montreal Protocol on protecting the ozone layer were implemented and emissions of chlorofluorocarbons (CFC) were cut but emissions of other greenhouse gases continued at their current rate, global temperature would rise by 0.3° centigrade per decade in the twenty-first century. To keep the decadal rise in temperature to 0.1° centigrade would require cutting fossil fuel emissions by 40 to 60 percent.[33] Limiting global warming to 0.1° centigrade a decade could only be accomplished with significant reductions in fossil fuel use, the report stressed.[34] In fact CO_2 emissions from fossil fuels have risen by 80 percent since then, but (decadally smoothed) average global temperature rose by only 0.1° centigrade a decade in the first 14 years of the current century.[35]

Even so, it is in the economics of climate change that the Villach/Bellagio group's bias is most evident. Adapting to climate change or cutting emissions could involve "high costs" to global society. Here the workshop encountered a problem. The costs of tackling climate change are incurred in the near future, but the benefits are far distant. To deal with this, the group came up with a similar solution to that advocated by Gottfried Feder and Franz Lawaczeck in the 1920s and 1930s: Abolish the rate of interest. "A major problem is the current practice of 'discounting' the future, since it is inappropriate to discount into present monetary values the risk of major transformations to the world of future generations," the group concluded.*[36] The logic of this argument requires that climate change be potentially catastrophic. Thus the catastrophizing of global warming—irrespective of the sci-

* A discount rate is an interest rate applied to future values. The higher the discount rate, the lower the present value of these. The valuation of time and its application to the economics of climate change are discussed further in Chapters 24 and 25 of Rupert Darwall, *The Age of Global Warming* (London, 2013).

entific evidence—is the logical consequence of the need to justify the preferred policy option—namely, steep cuts in fossil fuel consumption.

On the group's action list was consideration and development of the report's recommendations for the World Conference on the Changing Atmosphere (Toronto, June 1988) and the Second World Climate Conference (spring 1990, but in the end pushed back to November 1990 to coincide with the IPCC's First Assessment Report).[37] The Toronto climate conference was opened by Canadian premier Brian Mulroney and Gro Harlem Brundtland. A "coalition of reason" was required to bring about a rapid reduction of North–South inequalities and East–West tensions, which the conference said was a prerequisite for negotiating the agreements to secure a sustainable future for the planet.[38] With the threat of war between the nuclear superpowers rapidly receding, another threat was taking its place: "Humanity is conducting an unintended, uncontrolled, globally pervasive experiment whose ultimate consequences could be second only to a global nuclear war," the conference statement declared.[39] It was an absurd claim. The comparison with nuclear war intellectually and morally diminished those who had been pushing it. But it was necessary, for the reason revealed by Bolin's group the previous autumn, to reverse engineer a catastrophe.

How genuine was Sweden about global warming? Sweden has large amounts of peat and little coal. The carbon intensity of peat is higher than that of coal. On February 26, 1986, Palme was in central Sweden visiting a new peat-burning power station. It was the best moment of his tour, Palme told journalists.[40] Walking home from the cinema two days later, Palme was struck by a single assassin's bullet.

How does one fit the "overwhelming paradox" of Olof Palme and, indeed, the whole policy of the period? Theutenberg asked in his diary when he heard the news on the radio the next morning.

> Does it fit all together? A gigantic Machiavellian "double play"? A balance on the knife edge! Which in the end "toppled overboard" that cold evening on 28 February at Sveavägen? Somebody at last found "the double play" all too grave and gigantic? And put a stop to it![41]

A drug addict was convicted of the murder but acquitted on appeal.

The Social Democrats gave Palme a socialist funeral in Stockholm's Town Hall: the white coffin covered in red roses beneath red flags and a large placard displaying the emblem of the United Nations. The chairman of the Social Democratic Youth League and future foreign minister, Anna Lindh, who would die from an assassin's knife seventeen years later, gave a eulogy. Another was by someone twice touched by assassination. "In the truest sense, he belonged to all of us," Senator Ted Kennedy told mourners,

> and today I regard him as a brother. And, if I may be permitted, I would apply to him now some words I spoke for Robert Kennedy. Olof Palme saw war and tried to stop it. Let us pray that what he was for us and what he wished for others will some day come to pass for all the world.[42]

In the early years of global warming, there was far more skepticism and balance than would be permitted later on. A report in the *Financial Times* in 1990 on the IPCC's First Assessment Report ran beneath the subheadline "The evidence is much weaker than many pundits say." One IPCC contributor criticized the assessment report summary for not accurately reflecting the scientific discussions. "In the scientific papers, a great deal of care was devoted to pointing out the uncertainties," Andrew Solow of the Woods Hole Oceanographic Institution told the *Financial Times*'s David Thomas. It was a caricature to portray the climate debate as between an overwhelming majority committed to the position outlined in the policymakers' summary and a handful of mainly American-based scientists who regarded it as nonsense. According to Thomas, "A third, and large, group of scientists in the middle fears that the hype and the political pressures have pushed the scientific community beyond the bounds of the evidence."[43]

A 1992 Gallup survey of 400 U.S. experts conducted on behalf of the Washington, D.C.-based Center for Science, Technology & Media confirms this picture. While 60 percent of those surveyed believed that global average temperatures had increased over the past century, only 19 percent attributed this to human activities. Sixty-six percent believed that human-induced global warming was under way but only

41 percent believed that current scientific evidence supported this, and 70 percent rated the media coverage as "fair" to "poor." "By following national media coverage, one would not gain an accurate view of the scientific debate over global warming," the center's Mark Mills commented.[44]

The IPCC's First Assessment Report had been rushed to provide a scientific imprimatur to the planned Ministerial Declaration directly following the Second World Climate Conference in Geneva. This had required a preparatory meeting several months beforehand to agree a draft text. William Kininmonth, head of Australia's National Climate Center, attended the preparatory meeting and the conference. The Australian delegation to the preparatory meeting was led by foreign affairs officials supported by representatives from environment, energy, and industry departments but was to have no science representation. After strong representations from Zillman, Kininmonth was offered a place as the token scientist as long as the meteorological budget would pay for him to go. "The whole activity was orchestrated to produce a particular outcome: a Ministerial Declaration recommending the UN take action to produce a treaty to reduce CO_2 emissions," Kininmonth recalls.[45] Without actually knowing the outcome of the IPCC report (only the Summary for Policymakers of Working Group I had been circulated to governments), it was generally agreed (with square brackets to indicate provisional text) that the Ministerial Declaration should be for the UN to agree to negotiate a treaty to prevent dangerous climate change.

Copies of the IPCC report were distributed to delegates early in the last week of the conference itself. As it was closing, there was a motion from the podium that the report be accepted. "I doubt whether more than one in twenty had read the voluminous report in the time but nevertheless the report was accepted on a show of hands," Kininmonth says.[46] Detailed differences were settled in back-room discussions. The Ministerial Declaration was then agreed and sent to the UN. The science would continue to be taken care of in the capable hands of Professor Bolin, chosen by consensus to be the IPCC's first chair in reflection, records Zillman, of his standing as a world leader in climate change.

12 Sun Worship

Sunrise—a magical moment . . . I close my eyes. Once again the sun is rising over the Earth—as every morning for the last four and a half billion years. The more intently we observe the sun, the more she surprises us. This is what I feel now we have solar installations on our roof.

Franz Alt, "On the sunny side. Why the Energiewende *will make us all winners," 2013*[1]

The sun's rays deliver to our globe daily fifteen thousand times more energy than the daily consumption of nuclear and fossil energy. What we need is a mobilization of technologies to harvest this energy potential.

Hermann Scheer, 2008[2]

To have set out to fight Nazism in its sundry modern democratic guises, only to have ended up, in a modern left-wing guise, Nazi-like!

Paul Berman[3]

The politics of global warming originated in Sweden as a tool to promote nuclear power. Using wind and solar power to combat global warming originated in Germany, where the Greens had become a political force, thanks to popular opposition to nuclear power. Wind and solar did not emerge out of the logic of emissions cuts and hardly featured during global warming's first decade. A 20 percent cut in global emissions of greenhouse gas emissions by 2005 was what the Toronto climate conference said in 1988 was needed. Energy efficiency (i.e., cutting demand) could achieve half the emissions target. Industrialized nations could achieve the other half by "revisiting the nuclear power option," and advanced biomass conversion technologies could be deployed in the developing world, something that would encourage reforestation.[4] Wind and solar were not mentioned in the final conference document. It would take the best part of two decades for wind and solar to become de rigueur as the principal tool of decarbonization, a function they are ill-equipped to perform.

The Toronto climate conference had set out a general approach: "to internalize externalized costs and thereby consider the costs of the energy systems in their broadest sense."[5] "Here's a hunch," *The Economist* editorialized in May 1989.

> Within the next half-century the governments of many industrial countries will raise perhaps one-fifth of their revenues from taxes and charges on pollution. Largest of all will be a tax on the carbon dioxide when fossil fuels—coal, gas and oil—are burnt. . . . The sooner voters are willing to face the true costs of cutting carbon, the sooner will the world stop warming.[6]

It didn't quite turn out this way. Sweden, Norway, and the Netherlands implemented various forms of carbon taxation, and the European Commission championed one in the early 1990s. In the 1992 presidential election, Bill Clinton campaigned on Al Gore's BTU tax, which the Senate blocked in 1993, to Clinton's great relief.[7]

Governments soon soured on making the costs of decarbonizing excessively clear to voters. Instead of taxing emissions, the focus switched to capping them. In 2003, the European Union decided to implement cap-and-trade with its Emissions Trading Scheme (ETS). Just two years after the start of ETS in 2007, the EU adopted mandatory renewables targets and wrecked the ability of the ETS to cut emissions. Germany's renewables policy became European policy. Of their own accord, the rest of the world would fall into line. Australia followed with its own targets. Renewables were to become central to the Obama Administration's Clean Power Plan. America's climate agreements with China (2014) and Brazil (2015) also specify targets for renewable energy. Why did this happen? It is a story of the rise to power of the German Greens bringing about the greening of Europe and the subsequent spread of the green gospel around the world.

In 1989, nine years before coming to power in Berlin's first Red–Green coalition, the Greens' leader, Joschka Fischer, wrote a book of "radical ecological pragmatism" on restructuring industrial society.[8] Energy was most fundamental of all. An ecological energy system would "be the central innovation for an ecological restructuring of society."[9] A class of green rent seekers and lobbyists should be created. Fischer wanted an "entrepreneurial left" to make profits from a steadily growing environmental sector of the economy.[10]

The 1998 election that brought Fischer and the Greens into the federal government was not, however, a mandate to turn Germany green. In part, Germany's *Energiewende*—normally translated as Energy Transition, though *Wende* implies something big and significant*—was the product of a specific set of circumstances. In part, it was the outcome of a small group acting more like a revolutionary cell than a democratic party to bring about the most far-reaching changes to the electricity system since Edison first flicked the switch that brought electric light from the Pearl Street generator. Jobs had been the big issue of the election, and the big winner was the SPD, while the Greens polled slightly less than they had four years earlier. SPD leader Gerhard Schröder was a hard-headed pragmatist. The price of avoiding a grand coalition with the Christian Democratic Union (CDU) was doing a deal with the Greens.

The formation of the Red–Green coalition was historic. As Berman points out,

> under the Red-Green coalition, a greater number of veterans of the New Left had risen to power than in any other country among the big Western powers—risen through the Greens or else through the Social Democrats.[11]

Their arrival came at a time of crisis on the left. It was not, to be sure, anything like the New Left's existential crisis following the German Autumn 21 years earlier, but one that, temporarily weakening the Greens, made the push for renewable energy all the more necessary. The Greens had been formed as a political vehicle to oppose nuclear power and were defined by their policy of equidistance between the military blocs.

* *Die Wende* is the way Germans refer to the collapse of the Berlin Wall and reunification.

Four days before the 1998 elections, the UN Security Council passed Resolution 1199 on Kosovo. During the postelection transition, the outgoing and incoming governments united to back NATO intervention. "'Never again Auschwitz' is the historical admonition to prevent what could develop into a genocide," Fischer declared.[12] He carried his parliamentary caucus but, at a special convention of the Greens in May 1999, Fischer was pelted with a red paint bomb, bursting his ear drum.[13]

The Greens' East–West equidistance plank was no more, and Schröder wasn't going to let the Greens run up huge bills closing down nuclear power stations. Although the ex-communist Jürgen Trittin became environment minister in the coalition, the SPD refused to compensate the utility companies for shutting down their nuclear power stations. A June 2000 compromise implied extending the average operating life of the nuclear power stations by 32 years. The deal was promptly denounced by Friends of the Earth. With it went the credibility of the Greens' antinuclear plank. What was left of the causes that had called the Greens into being? If it didn't do something on renewables, the Greens would no longer be green.

Although a big renewables program was essential for the Greens, the landmark 2000 Renewable Energy Act, known in Germany as the EEG (standing for *Erneuerbare Energien Gesetz*) was conceived, crafted, and pushed into law by a small group of backbench Green and SPD MPs with the help of the top Green bureaucrat at the environment ministry. Indeed, it can be fairly said that Germany's *Energiewende* began with an *Energie Putsch*. According to the journalist Alexander Wendt, the law was never included in any party election campaign or manifesto. There was no public debate and no prior impact assessment. MPs voting on it had little idea what they were actually voting for. Large parts of the SPD were against the EEG, including the SPD energy minister, Werner Müller, a former energy utility executive, who was alarmed by its extremely generous feed-in tariffs for renewable energy.[14]

The leader of the putsch was Hermann Scheer. Acclaimed by the *Toronto Globe and Mail* as "probably the most influential renewable-energy lawmaker on the planet,"[15] Scheer was a radical, a revolutionary even. A Sixty-eighter, Scheer had been vice chairman of the SPD's youth wing before being elected to the Bundestag in 1980, aged 36. There he developed a special interest in peace and disarmament issues and aligned himself with the communist-backed Peace Movement. Scheer was a prophet who, unlike Moses, reached the promised land, one that flowed with milk and honey, not for the people but for a privileged few. Scheer the prophet was also a parasite, seamlessly combining the role of legislator with his activities as a lobbyist. In 1988 he established EUROSOLAR trade association with the aim of making the replacement of nuclear and fossil fuels the "task of the century" and in 2001 founded the World Council for Renewable Energy to lobby and propagandize at a global level.[16] He electrified a Washington audience of environmental activists in August 2008 with his gale-force enthusiasm, *Time* reported.[17] Some called Scheer "Europe's Al Gore," a comparison that does not do Scheer justice. As well as enriching himself and his family (EUROSOLAR employed his wife and daughter), Scheer actually accomplished something. "This is much more—much more—than only the exchange of technology," Scheer told a Canadian audience in 2008.

> This is a change of the whole energy economy. Automatically there is a change in the industrial development. It is a change in the cultural development. It is a change of international relations. Automatically. It is a change of financing structures, a change of the world finance system.[18]

Scheer's interest in renewable energy developed out of his activities in the Peace Movement. In the mid-1980s, Scheer had written a book on disarmament, *Liberation from the Bomb.* A more accurate title would be *Let's Surrender to the Soviet Union*. Scheer argued that Western Europe should liberate itself from American influence and join with Eastern Europe in a neutral European Socialist Federation.[19] It would then develop close links with developing countries; European industries should stop trying to compete in pointless competition with the United States and Japan in high-tech markets but fulfill the basic needs of people in the Third World. The defense industry would be closed down and its engineers and scientists redirected to creating a new energy paradigm based on solar and other forms of renewable energy, to transform society from its dependence on fossil fuels. By harvesting the sun from the Sahara, Europe would be transformed into a hydrogen society.[20] *New Scientist* asked Scheer how, without any background as a physicist or engineer, he had gotten involved in solar energy. It was the time of Ronald Reagan and his Strategic Defense Initiative. "I had not read a single book on renewable energy. I just did my own thinking and I wrote a chapter suggesting a new SDI, the Solar Development Initiative," Scheer replied.[21]

Unsurprisingly Scheer found little to celebrate in the ending of the Cold War. In *A Solar Manifesto,* published four years after the fall of the Berlin Wall, Scheer condemned the leaders of the West for "their self-deceiving euphoria of victory."[22] Even so, Scheer managed to profit from this apparent reverse. Intended as a postreunification boost to small-scale hydro schemes, the 1991 Electricity Feed-In Act had Scheer's fingerprints all over it. It rewarded the most inefficient generating technologies with a sliding scale of guaranteed prices for privately generated electricity (99 pfennigs [60¢] per kWh for solar PV down to 15 pfennigs [9¢] per kWh for small hydro schemes). The 1991 Feed-In Act was a charter for rent seekers. Later he would boast how EUROSOLAR had got the European Commission to propose a similar structure as the basis of a 1997 commission white paper.[23]

If Fischer was an exponent of radical ecological pragmatism, Scheer was more into the revolutionary variety. He quoted approvingly the claim of the German writer Carl Amery that mankind's survival dictated the "fastest possible destruction of the industrial system, at any price." Amery's 1976 *Nature as Politics* had provided a theoretical justification for future Green policies that placed capitalist industry as the final stage before a postindustrial green economy. A leftist, pacifist, and member of the SPD before joining the Greens, Amery (an anagram of the Munich-based Christian Mayer) was the first serious left-wing writer in West Germany to develop a materialistic (i.e., Marxian) theory for a nature-centered political philosophy. Taking a different route from the metaphysical philosophizing of the German anticapitalist right, the New Left and the Old Right ended up in the same place: Red and Brown become Green. A further facet of Red–Brown convergence was in the writings of the post–Frankfurt School theorist Robert Kurz, another author name-checked by Scheer. Nothing short of "the genuine abolition of modern goods and their global system" would be enough to avoid global suicide.[24] The overthrow of the American-led global economic order was an aim Nazis and the New Left had in common.

According to Scheer, energy policy was the lever to transform society. Social development was a function of economic development that was in turn determined by access to energy. "It is no accident that the history of human development has always been a history of the various energy supply options," Scheer wrote.[25] Replacing hydrocarbons and nuclear energy would close mankind's breach with nature. The society of the future was to be found in the sun: "Humanity has the chance to survive only if it is able, within a short period of time, to replace conventional energy sources with the solar energy that flows through the planet's ecological system."[26] The solar revolution would be even more important than the French and Industrial Revolutions and be "the decisive step towards incorporating humanity into the 'rhythm of nature,'" Scheer wrote.[27]

There was no slackening, let alone reversal, in the wind and solar revolution after the fall of the Red–Green coalition in 2005. Within two years of becoming chancellor, Angela Merkel had induced the European Union to adopt a 20 percent mandatory renewables target. Merkel understood the perverse effect of imposing a renewables mandate on cap and trade. More wind and solar electricity in Germany means German power companies need fewer emissions allowances, depressing their price so Spanish and Italian power companies can burn more coal. According to Fritz Vahrenholt, who was environment minister in Hamburg when Merkel was federal environment minister in the Kohl government, the future chancellor had been strongly against renewables and recognized that the effect of German wind farms would be to simply push emissions to other EU member states.[28] As late as June 2005, Merkel had declared that "increasing the share played by renewable energy and electricity consumption to twenty percent is hardly realistic."[29] Yet two years later, not only had she signed up Germany to a renewables target, she'd got the whole of the European Union on the hook of a 20 percent renewables target.

What explains Merkel's 180-degree turn? The September 2005 federal elections had been Merkel's to lose. Having pushed through unpopular labor market and welfare reforms, Schröder's SPD was divided and exhausted. After a string of regional election defeats, Schröder contrived a parliamentary defeat to trigger early elections. A poll a week after the parliamentary vote showed Merkel's CDU and its Bavarian Christian Social Union (CSU) sister party with a 22 percentage point lead over the SPD (47 percent to 25 percent). Merkel proceeded to fritter it away.[30] On polling day, her lead had shrunk to one percentage point in the second-worst result for the CDU/CSU since 1949. After nearly two months of postelection haggling, the outcome was the first CDU/SPD grand coalition since 1969, with Merkel as chancellor.

Merkel's near-death experience turned her into a political calculating machine. Ten years on, Merkel remained a slave to opinion polls —reportedly commissioning over 600 of them between 2009 and 2013.[31] Her adoption of the renewables target, Vahrenholt says, is like an ice cream seller on a crowded beach who knows she can sell more ice creams by starting at the middle and squeezing the competition.[32] It was a masterstroke. The renewables target put the SPD in an impossible position. It split its blue-collar base from the party's public-sector, white-collar professional demographic that strongly identifies with ecological values, at the same time isolating the SPD's regional stronghold of North Rhine-Westphalia, known historically as the Coal and Steel State.

Merkel then used Germany's presidency of the European Council in 2007 to get European leaders to agree that the EU's 2020 climate and energy package should include a binding commitment for the EU to derive 20 percent of its energy from renewable sources by 2020. In London, ministers and officials were aghast that Tony Blair, ever the sucker for the grand gesture, had committed Britain to the most costly renewables target of any EU member state. A 19-page paper prepared by civil servants was sent to ministers reviewing possible options to contain the damage. They noted that the renewables commitment risked making the ETS redundant and would cause the price of emissions allowances to collapse. The renewables target would triple the cost of meeting the United Kingdom's emissions target compared to the flexibility of emissions trading, the officials estimated. In canvassing potential allies (Italy looked the best hope), the paper recognized that Germany was the principal stumbling block: there was the politics of nuclear power, the prospects for Germany's renewables industry (a bonanza that was to materialize in China), and the fact that "Merkel personally championed it at [the] Spring Council."[33] Their efforts were to no avail. Britain was locked in.

So was Germany. By 2013, Germany had installed more solar capacity than any other nation, much of it imported from China. The irreversibility of Scheer's *Energiewende* was reflected in hard investment

Table 1: Electrical generating capacity and output in Germany—2013 (preliminary data)

	Capacity		Output	
	GW	%	TWh	%
Solar PV	36.3	19.2%	31.0	4.9%
Wind	34.7	18.3%	51.7	8.2%
Hard coal	29.2	15.4%	121.7	19.2%
Natural gas	26.7	14.1%	67.5	10.7%
Lignite	23.1	12.2%	160.9	25.4%
Nuclear	12.1	6.4%	97.3	15.4%
Hydro	10.3	5.4%	28.8	4.5%
Other	17.0	9.0%	74.3	11.7%
Total	189.4	100.0%	633.2	100.0%

Source: Federal Ministry for Economic Affairs and Energy, "Energy Data: Complete Edition" (last updated October 21, 2014), Table 22 http://www.bmwi.de/EN/Topics/Energy/Energy-data-and-forecasts/energy-data.html (accessed September 7, 2015).

in Germany's electricity generating mix: By 2013, solar photovoltaic (PV) panels and wind turbines represented 37.5 percent of total generating capacity. In fact, windy, cloudy, Germany had more solar (19.2 percent of its electrical generating capacity) than wind (18.3 percent), which was its single largest chunk of generating capacity (Table 1). But solar PV was even less efficient than wind. Although there was 4.6 percent more solar PV capacity than wind, it generated 40 percent less electricity. Overall, solar and wind accounted for three-eighths of installed generating capacity but generated slightly more than one-eighth (13.1 percent) of Germany's electricity. Power stations burning lignite, a type of coal closer to peat than to hard coal, accounted for less than one-eighth of generating capacity but produced one-fourth of Germany's electricity. It meant that in 2013, green Germany derived over five times as much electricity from a fuel demonized by environmentalists than it did from solar PV.

In Europe, 10 percent of the popular vote was "a mandate to change the world," Brookings's William Antholis had told the 2009 Essen Great Transformation conference.[34] As we've seen, Germany's 1998 election hadn't been a mandate to turn Germany green. Nor could it possibly change Germany's geography so it could get more sun. Bavaria is not Arizona. Germany's southernmost point is just north of the 47th parallel, which in North America runs through Maine, Quebec, and Ontario and across to the state of Washington. At 55 degrees north, Schleswig-Holstein's northern tip is on the same latitude as Alaska and Siberia. Worse than the latitude is the weather. Western Europe's climate is dominated by weather systems from the North Atlantic, so it has much higher levels of cloud cover than equivalent latitudes in North America. A solar PV panel in Phoenix, Arizona, can be expected to generate 75 percent more electricity than one in Munich, Bavaria. A PV panel in Hamburg is likely to generate half the electricity of one in Abu Dhabi.[35]

What was it about Germany and the sun? Reason does not supply an answer as to why Germany became the country with the most installed solar PV capacity in the world. The explanation lies elsewhere, in German culture and thought. Nietszche's Zarathustra stepped out of his cave to stand before the sun: "Great star! What would your happiness be, if you had not those for whom you shine."[36] A clue is Scheer's praise for the Nobel Prize–winning chemist Wilhelm Ostwald.* In a 1909 book on natural science and philosophy, Ostwald wrote that the sun sends us "free energy" that drives practically everything that happens on earth. As summarized by Scheer, Ostwald surmised that "an enduring economy must be based exclusively on the regular utilization of the annual solar radiation energy."[37] This, of course, is pure claptrap. Just because the sun is necessary for life doesn't mean that the sun's energy is responsible for everything; there are subsidiary causes and there are things other than the sun that are also essential for life—gravity, for example, the Earth, its atmosphere, and the oceans. But none of these account for humanity's progress. They existed before the first humans and will exist after the last. What was Ostwald's logic in singling out the sun? In coming close to the Aristotelian notion of the "prime mover," which medieval theologians adduced as proof for the existence of God, Ostwald's solar philosophy teeters on the brink of paganism.

* Ostwald's scientific work was greatly influenced by Svante Arrhenius. He was awarded his Nobel in 1909, six years after his Swedish colleague.

Ostwald makes an appearance in the chapter on the socialist roots of Nazism in *The Road to Serfdom*. Hayek argued that it had been the union of the anticapitalist forces of the right and the left, "the fusion of radical and conservative socialism," that had driven everything that was liberal out of Germany.[38] In Ostwald's system, his energy ideas form the basis for the collectivist organization of society. Society was like a well-functioning body in which individuals must submit to the imperative of maximizing the energy efficiency of the whole:

> I will explain to you now Germany's great secret: we, or perhaps the German race, have discovered the significance of organization. While the other nations still live under the regime of individualism, we have already achieved that of organization.[39]

As well as collectivism, the recurring motif of the sun also connects modern-day Greens to the pre-1945 era. Sun worship had been a feature of early Nazi racial ideology. "Sometimes it looks as though a new sun is about to rise in the south," Joseph Goebbels wrote in 1922.[40] The early Greens' sunflower logo was designed by the avant-garde artist and Nazi Joseph Beuys. It had its first major public outing at a Green Party conference in Frankfurt in 1979, when August Haussleiter and the Far Right AUD constituted the dominant faction. Beuys was also a member of the AUD, standing as an AUD candidate in the 1976 federal elections. Beuys did not hide his membership in the Hitler Youth and suffered severe depression following Hitler's suicide. He surrounded himself with former and long-time Nazis, who were his artistic patrons and his political comrades-in-arms.[41] As for the Nazis' logo, Beuys would surely have known that the swastika was originally an Indian sun symbol.

Antagonism to industrialization is a recurrent and irresolvable contradiction of German culture. Otto von Bismarck had united Germany with blood and iron. Coal and steel propelled Germany's ascendancy in Europe. Germany's industries constitute the economic basis for the country's leadership of Europe. Limiting energy usage each year to the amount of energy received from the sun would have precluded the Industrial Revolution, which was powered by releasing fossilized solar energy in the form of coal. The world's second-largest exporter of manufactured goods cannot depend on wind and solar for its energy. This is not a dilemma for the Greens, as they want to reverse the

Industrial Revolution, but what about Angela Merkel? The answer to this part of the Merkel enigma can be found in one of the most influential and least known of the leading figures in the Greens.

Described by colleagues as fierce, tough, and clever, Rainer Baake trained as an economist before spending four years as a community organizer in Chicago in the mid-1970s. After a stint as a local councilor, in 1991 he was appointed to the top bureaucratic position working for Joschka Fischer when he was environment minister in the state of Hesse. In 1998, Baake was appointed state secretary (the top permanent bureaucrat) at the federal environment ministry under Trittin, where he was responsible for the nuclear phase-out deal and the EEG renewable energy legislation. After the fall of the Red–Green coalition, Baake worked for federally funded environmental NGOs pushing for acceleration of the *Energiewende* and in 2008 became an environmental adviser to the CDU minister president of Hesse. After the 2013 federal elections, Merkel formed a second grand coalition with the SPD. Sigmar Gabriel, the new SPD leader, was appointed federal minister for economic affairs and energy. What did Merkel do? She appointed Baake state secretary to keep an eye on her chief political rival. As far as energy policy was concerned, Germany wasn't being governed by the CDU/SPD grand coalition but by a CDU–Green coalition. Not only was Germany locked into the *Energiewende*; with Baake's appointment, Merkel had thrown away the key.

13

Renewable Destruction

It remains the case that supporting renewable energy costs the average household about only one euro a month—the cost of a scoop of ice cream.

Jürgen Trittin (environment minister), 2004[1]

The costs of our energy reform and restructuring of energy provision could amount to around one trillion euros by the end of the 2030s.

Peter Altmaier (environment minister), 2013[2]

Like their forebears, Berlin's green revolutionaries knew what they wanted to destroy but had little idea how to make their Utopia work. The *Energiewende* was full of surprises and unintended consequences. Germany was going to export wind turbines and solar PV panels around the globe. Instead the 2000 EEG renewable energy law spawned 100,000 profiteers and a gigantic solar industry in China.[3] Solar PV prices collapsed, and soaring solar take-up blew forecasts out of the water as the solar industry successfully lobbied against cutbacks to already excessive solar feed-in tariffs.

Having too much wind and solar meant that on summer days when the sun shone and the wind blew, more electricity was being produced than anyone knew what to do with. Not that it greatly helped cut carbon dioxide emissions. The big fall in German power station emissions had happened earlier, when many lignite-burning power stations in East Germany were closed after reunification. Between 1990 and 1999, power station emissions fell by 80.5 million metric tons from 423.4 million metric tons, a fall of 19 percent (Figure 1). After 1999, the only two consecutive years of falling power station CO_2 emissions were in 2008 and 2009, when German industrial output slumped by

Figure 1: Germany—Installed Wind and Solar Capacity (GW) and annual energy sector CO_2 emissions (million metric tons)

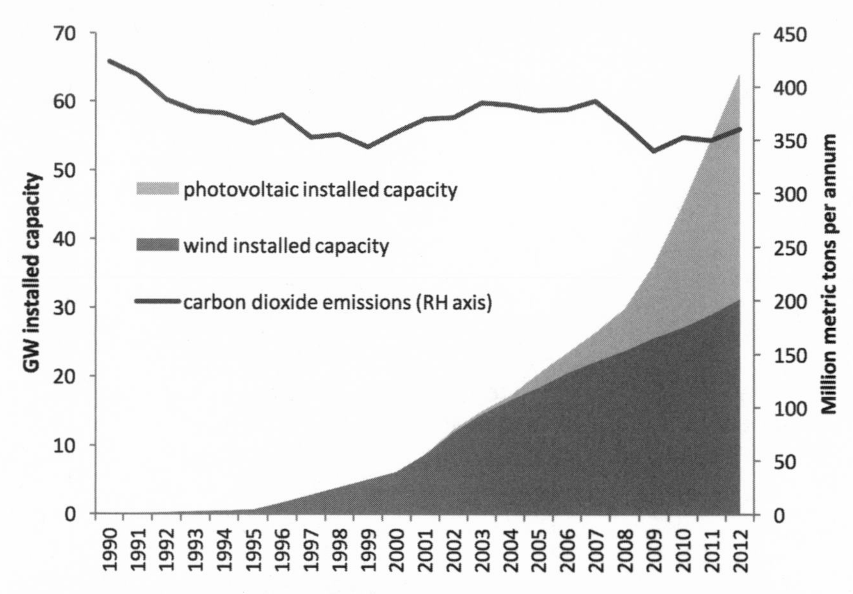

Sources: Federal Ministry of Economic Affairs and Energy, "Development of renewable energy sources in Germany 2014," http://www.erneuerbare-energien.de/EE/Redaktion/DE/Downloads/development-of-renewable-energy-sources-in-germany-2014.pdf?__blob=publicationFile&v=6 (accessed July 4, 2015); Eionet, Central Data Repository, http://cdr.eionet.europa.eu/de/eu/ghgmm/envutt6ka (accessed July 4, 2015).

more than 20 percent. Second only to reunification, the subprime crisis and Lehman bankruptcy turned out to be Germany's most effective decarbonization policy, though scarcely intended as such. By 2012, CO_2 emissions from power stations were 17.2 million metric tons higher than in 1999—an increase of 5 percent. Over the same period, installed wind and solar PV capacity rose nearly 13-fold—from 5 GW in 1999 to 64.3 GW in 2012.

"All of us under-estimated this legislation," admitted Michaele Hustedt, a Green MP and cosponsor of the EEG law.[4] The SPD energy minister Werner Müller didn't, warning that the subsidy would be far too generous. To head off the threat, Müller wanted to establish a facility in southern Europe to research solar energy. As he feared, the rush to wind and solar quickly pushed up electricity bills. As shown in the Renewable Hockey Stick (Figure 2), at low levels of wind and solar PV penetration, renewables have little discernible impact on

Figure 2: Renewable Hockey Stick: Comparison of wind and solar capacity and electricity prices, selected European countries (2012)

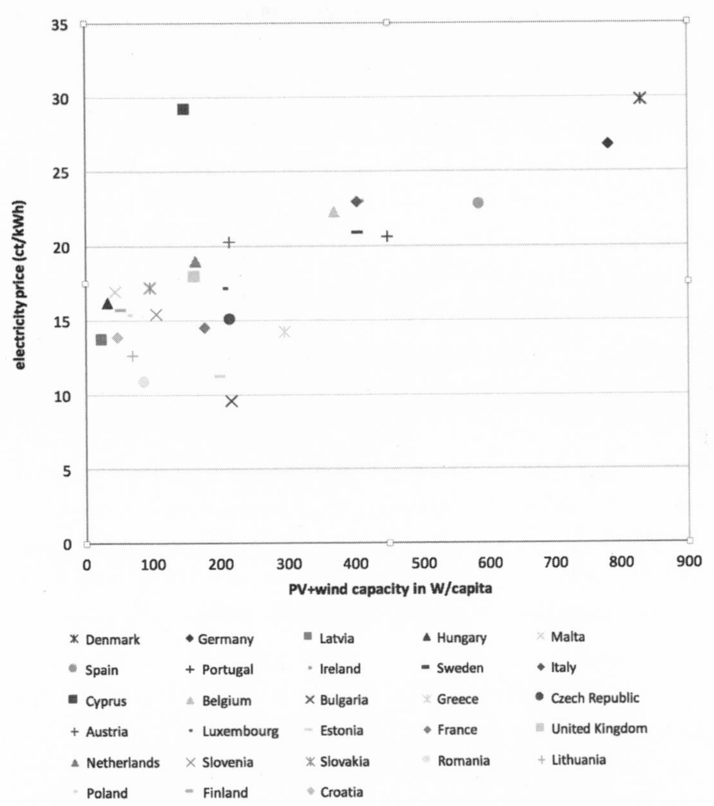

Source: Hans Poser, Jeffrey Altman, Felix ab Egg, and Andreas Granata, Ross Board, *Development and Integration of Renewable Energy: Lessons Learned from Germany* (July 2014), Figure 16, www.finadvice.ch.

prices. Up to around 200 watts of wind and solar capacity per capita (the hockey stick blade), electricity prices cluster around 15 euro cents (18¢) per kWh. Above 200 watts comes the shaft. As more renewables are put on the grid, prices rise. Including taxes, Danes, with the highest proportion of wind capacity in the world, also had the highest electricity prices in the world, at 29.72 euro cents (36.7¢) per kWh. They were closely followed by Germans, who paid 26.76 euro cents (33.1¢). Americans, with combined wind and solar PV capacity of 200 watts per capita, paid an average price of 7.5 euro cents (9.3¢) per kWh.[5] In 2012, thanks mainly to the headlong rush to wind and solar, Danes paid four times and Germans three and a half times what Americans did for their electricity.

Third-place Spain, like Germany, was hit by the double whammy of falling solar PV prices and gold-plated subsidies paying out 12 times the market price. "Sustainable energy meets unsustainable costs," *The Economist* reported in 2012 on the Spanish "cost del sol." In 2007, Spain had 690 MW of installed solar capacity; five years later, it was more than 6 GW (though less than one-fifth of Germany's solar capacity) and subsidies to all renewables had risen eightfold to €8.1 billion a year—nearly one percent of Spanish GDP. To hide the costs, the Madrid government stopped energy utilities passing on the full cost of the price supports. Instead they ran up a tariff deficit at the rate of €5 billion a year to reach a cumulative deficit of €26 billion in 2012, adding to Spain's fiscal problems. When the euro crisis struck, instead of raising electricity prices, the Spanish government retroactively cut renewable subsidies. It had failed to cut subsidies when renewables were booming, so the cuts had to be draconian. The promised green jobs vanished. "It has been a chastening experience," *The Economist* concluded.[6] That was one way of putting it.

Britain, a latecomer to the wind and solar party, also experienced skyrocketing electricity prices. In the ten years from 2004, average domestic electricity prices increased by 75 percent. When allowing for its low rate of tax on fuel and power, Britain's domestic electricity prices were the second highest in the EU15 (i.e., excluding Cyprus and Eastern Europe).[7] Britain uses a portfolio standard. Electricity suppliers have to buy green Renewable Obligation Certificates (ROC). By 2020/21, ROC payments are projected to reach almost £4 billion ($6.3b) a year and add nearly nine percent to domestic electricity bills.[8] There are also generous feed-in tariffs to encourage households to put solar panels on their roofs. Within four years, cumulative spending of £2.1 billion ($3.3b) was nearly double the £1.1 billion ($1.7b) originally forecast.[9]

Looking for someone to blame other than themselves, British politicians accused energy companies of price gouging. After conducting a market investigation, the Competition and Markets Authority (CMA) confirmed that the main driver of domestic electricity price increases was not profiteering by energy companies but the cost of government-imposed obligations and network costs, the latter largely reflecting the costs of integrating wind and solar capacity. Gas and electricity take up nearly ten percent of the household spending of the poorest ten percent of the population and is their largest item of expenditure after housing. "This relationship between expenditure on energy and income explains part of the concern around energy price increases—they have a highly regressive impact," the CMA said.[10] The outlook is for the squeeze to intensify. Green levies are forecast to almost quadruple from £3.6 billion ($5.7b) in 2014/15 to £13.6 billion ($21.5b) in just five years.[11] It is the least well off who are hit hardest by green virtue.

Despite the demanding nature of Britain's commitment to cut greenhouse gas emissions from 1990 levels by 35 percent by 2020, the CMA found the renewables target "more constraining."[12] Using renewables inflicts additional economic pain for no societal gain, a finding with important implications for others following Germany's lead, including several states in the U.S., notably California. The CMA was also critical of having a renewables target in parallel with cap and trade:

> while the ETS cap is still in place and binding, such policies will not reduce aggregate EU or global emissions, but serve to depress the EUA [carbon credit] price.[13]

One after another, non-EU countries fell into that elephant trap. During his September 2015 state visit to the United States, President Xi Jinping announced that in addition to promoting renewable energy, China would also adopt cap and trade. In terms of policy smarts, the mandarins of the Middle Kingdom have little on their counterparts in Berlin, Brussels, and Washington.

Britain and Spain were also-rans in the race for the most painful way not to decarbonize. Unlike Spain, Germany did not have a fiscal crisis to provide a shock of cold reality. Its green lobby was too entrenched to prune back renewable subsidies, as David Cameron did after winning the 2015 British general election. The cost of Germany's victory was colossal. When the EEG law came into force, the Green environment minister Jürgen Trittin claimed it would cost the equivalent of a scoop of ice cream on monthly electricity bills. In 2015, cumulative feed-in subsidies paid by consumers for wind and solar electricity had spiraled to €82.5 billion, and the accrued cost of *Energiewende* had reached €400 billion.[14] It turned out to be a mighty expensive ice cream. Nine years later, Trittin's successor as environment minister, the CDU's Peter Altmaier, reckoned the *Energiewende* could cost up to one trillion euros by the end of the 2030s.

None of this bothered Hermann Scheer and his fellow EEG putschists. The *Energiewende* had created "irrevocable" structural change, Scheer exulted on the EEG's tenth anniversary. Questioned about its cost, Hustedt was dismissive. "This debate is dangerous and is created by enemies."[15] It mattered to voters, so the economics of renewables had to be sold on a false prospectus. As Wendt notes, "all the predicted costs, timetable and consequences of *Energiewende* turned out to be wrong."[16] In 2005, Sigmar Gabriel had said that EEG costs would level out.[17] They kept on rising. Five years later, the feed-in tariff subsidy had nearly tripled.[18] In 2011, Angela Merkel pledged that the EEG subsidy would not rise above 3.5 euro cents (4.3¢) per kilowatt hour. Even in Germany, things had to have gotten pretty bad for a politician to start making pledges in kilowatt hours. Two years later, the EEG subsidy reached 5.3 euro cents (6.5¢) per kWh—breaking Merkel's pledge by 50 percent.[19] The same thing happened on jobs. The trade unions and the SPD had been promised hundreds of thousands of green jobs.[20] German solar subsidies did create jobs—in China. By 2009, it was evident the green job mirage was just that, so the environment ministry dialed back on the *Energiewende* jobs bonanza. The goal of environment policy, it now stated, "is not primarily as many jobs as possible, but rather to reach environmental goals efficiently."[21] That too was false.

In fact, *Energiewende* isn't a policy; it is a fairy tale relying on make-believe. The Merkel government's plan to derive 80 percent of electricity from renewables within 40 years assumes that German electricity production will be cut by 45 percent compared to 2008, from 640 terawatt hours (TWh) in 2008 to 350 TWh. This, so the theory goes, is to be achieved by cutting electricity demand by one-fifth and by importing one-fourth of Germany's electricity needs.[22] Demand reduction is a deeply held tenet of green ideology defying reality. In 2013, German households consumed 21.15 billion kWh more electricity than in 1990, a near one-fifth increase.[23] Another piece of make-believe assumes that the weather in Germany will be different from next door. Quite how Germany can be a net importer of electricity when its neighbors—exposed to the same North Atlantic weather systems—are also increasing their dependence on weather-generated electricity is only one of the magical elements in the *Energiewende* fantasy.

The costs of the *Energiewende* fairy tale are all too real, and attempts to rein them in came to naught. Merkel made the problem worse with her panicked response to the Fukushima nuclear disaster in March 2011. Helmut Kohl, her CDU predecessor as chancellor, had to handle a worse PR crisis in the wake of the Chernobyl disaster, the only accident in the history of commercial nuclear power to cause fatalities from radiation. Japanese authorities estimate that the amount of radioactivity released at Fukushima was only ten percent of the amount released from Chernobyl.[24] In response, Kohl appointed a minister for nuclear safety. Merkel ordered the closure of the nine oldest nuclear reactors and accelerated the phase-out of the rest. Replacing zero-emission, non–weather dependent capacity with high-carbon lignite power might have been designed to increase Germany's carbon dioxide emissions.

The public backlash against rising electricity prices kept Merkel on the defensive. In May 2012, she fired her environment minister, Norbert Röttgen. "The energy revolution is a central plan of this term and the foundations for it have been laid but we still have a lot of work to do," Merkel declared.[25] He was replaced by Altmaier, one of her most trusted aides, who evangelized that increased electricity costs would make people prudent and published a pamphlet extolling the benefits of energy saving. "Avoid pre-heating and utilize residual heat," Altmaier wrote in the introduction. TV viewers could also save a lot of electricity, albeit at the expense of picture quality. "For instance, you can reduce brightness and contrast," the booklet suggested.[26]

Throughout, Merkel's reaction to the consequences of her own energy policies was entirely political. After the 2013 elections, when SPD leader Sigmar Gabriel questioned *Energiewende* as vice chancellor of a second grand coalition, he was immediately attacked by his own party in Schleswig-Holstein, where wind vested interests are strong. This precipitated a fight within the federal government and between states, some wanting to cut feed-in tariffs, others wanting to keep them. Without Merkel's backing and with the Green Rainer Baake running his ministry, Gabriel never stood a chance.

How did Germany get it so wrong? According to Fritz Vahrenholt, the problems created by *Energiewende* were not foreseen. Not fighting *Energiewende* had been a "huge failure" on the part of the trades unions and something they came to bitterly regret. IG Metall, Germany's largest union, had been conned by the promise of 400,000 green jobs that never materialized. Although the SPD had little interest in saving the climate, it too had been captured by the false jobs promise.[27] Business was also complacent. Energy-intensive firms were bought off with 90 percent exemptions from EEG levies. Even SolarWorld, a leading manufacturer and marketer of solar PV, demanded an exemption or else it would leave Germany because of uncompetitive energy costs.[28] Germany's Big Four utility companies were anesthetized with bribes of grandfathered carbon credits before their business model was euthanized. RWE, the second largest, received a windfall of about $6.4 billion in the first three years of the ETS.[29] From their peak in January 2008, the three quoted utilities saw their share prices capitulate before the *Energiewende* machine of destruction.* By August 2015, the share price of EnBW had lost 59.5 percent of its value, e.on had fallen by 70.2 percent, and RWE by 84.7 percent. Over the same period, the DAX 30 index of German blue-chip companies had risen by 39.2 percent.[30]

This was not a market-driven, Schumpeterian gale of creative destruction. When markets destroy older businesses, it's because newer ones come along offering customers newer and better products and lower prices. By contrast, the *Energiewende* value destruction was entirely driven by the state. Its scale was immense. In the seven years from December 2007, the three quoted utility companies saw €68.9 billion of shareholder value destroyed.

This wasn't because households were paying lower electricity bills. Assuming Germany's 40 million households consumed an average 3,500 kWh a year, the total amount they paid for electricity increased from €27.2 billion in 2006 to €40.3 billion in 2015—a near 50 percent rise in nine years.[31] Indeed, the value loss to customers is a multiple of that inflicted on investors. If households have a 5 percent discount rate, it implies the capitalized cost of the higher electricity

* The fourth, Vattenfall, is 100 percent owned by the Swedish state.

price they paid in 2015 over the 2006 price—before taking account of further prices rises in the future—is €262 billion, nearly four times the shareholder losses of the Big Four utility companies.[32] Although early green investors made fortunes and *Energiewende* created a green Junker class of landowners receiving €40,000 ($49,000) a year for a large wind tower, many later investors were not so fortunate. A 2014 survey of private-equity renewables investments found that only 22 percent of renewables funds earned an internal rate of return higher than 3 percent. An investment adviser called renewable energy investing a "trendy fad" overly reliant on tax credits and subsidies. "One is investing more in political decisions than in tangible assets," Jay Yoder of Altius Associates told the *Financial Times*.[33]

Neither were consumers getting "better" electricity in return for higher prices. Although the cost of feed-in tariffs caught the public's attention, hidden from public view were spiraling grid costs to connect wind and solar farms scattered over the countryside to businesses and homes. Wind farms in the north of Germany need grid reinforcement and new transmission lines to the south. In just four years, the revenues charged by Germany's four grid companies jumped by nearly two-thirds, from €16.33 billion ($20.1b) in 2010 to €26.91 billion ($33.3b) in 2014.[34] Germany's grid had been the most reliable in the world, with a reliability of 99.95 percent. German engineers had built in so much redundancy into the grid that brownouts had been avoided, and until 2008 there had been no emergency grid interventions. In 2012 there were 1,000 brownouts, and in 2013 over 2,500.[35] A 2012 survey of members of the Association of German Industrial Energy Companies found that the number of short interruptions had grown by 29 percent in three years. Large and small companies were buying generators to protect themselves from a less reliable grid.[36]

While German consumers pay the second-highest electricity prices in Europe, on sunny, windy days, there is too much electricity. Streetlights and stadium lights are switched on in the middle of the day. Electricity that can't be burned off is dumped. Germany pays Switzerland and the Netherlands €100 million to 200 million a year to take away its unwanted renewable energy. To protect its infrastructure from German "garbage power," the EU gave Poland permission to build inverters to stop German electricity flowing over the Oder.

Because the ETS put a hard cap on EU-wide emissions, the EEG merely shifted carbon dioxide emissions around the EU. A 2009 study found that the EEG increased the profits of Italian and Spanish coal-burning utilities, boosting the profits of Enel and Endesa by 9 percent and 16 percent, respectively.[37] Overall, the paper came to a damning verdict on Germany's renewables policy: "the EEG's net climate effect has been equal to zero."[38]

The most lasting damage caused by Germany's failed climate policies is likely to be on the German economy. The EEG exemption was structured to benefit large energy consumers. This put the vast majority of small- and medium-sized businesses—the famed German *Mittelstand,* contributing to over half the nation's total economic output and generating two trillion euros of annual revenues—on the hook for the rising costs of *Energiewende.*[39] In the eight years to 2014, electricity prices for businesses consuming between 160 to 120,000 megawatt hours (MWh) a year increased by 35 percent, a lower rise than for households, as they did not have to foot the bill for exempting the largest electricity consumers.[40] The value of the 90 percent exemption created a perverse incentive to overconsume electricity to keep above the threshold. Unsurprisingly, the number of exempt firms has been rising. Funding the exemption constitutes around one-fourth of the household levy. Big energy users' gain is households' loss, setting up a conflict between households and industry in which the biggest winner is the Greens' deindustrialization agenda.

Despite the exemptions, a January 2014 research note by Deutsche Bank found that Germany's energy cost penalty was already eroding its industrial base. German industrial users paid 26 percent more for electricity compared to the EU average, while the disparity with the United States was even more pronounced.[41] In only two of the previous 17 years had companies in the energy-intensive sectors invested in excess of depreciation, indicating they were running down their assets.[42] At the same time, investment by German companies abroad had been increasing. A 2013 survey by the German Chambers of Commerce found that nearly one-third of all German companies and over half of industrial companies reported that *Energiewende* was having a negative or very negative impact on their competitiveness.[43] To restore German competitiveness, the Deutsche Bank note argued that Germany had to either get international agreement on more stringent global decarbonization targets and press its European partners to catch up with Germany—revealing an underlying economic motivation for German displays of eco-virtue at climate change negotiating fora—or decelerate the *Energiewende*.[44]

A year later, Deutsche Bank renewed its warning. Investment had not kept pace with the growth in output. As a result, Germany's industrial capital stock was being worn out.[45] The EEG exemptions were not a cure because companies were not confident that they would be sustained. "Major German chemical companies have explicitly cited energy policy in the United States as the main reason for their decision to increase capital spending in that country," the note continued.

> Germany's pursuit of unilateral policies on energy and climate change could encourage investment leakage and carbon leakage. A further problem is that lack of investment in energy-intensive sectors indirectly impacts on downstream industries in Germany. There are very few signs that energy-intensive companies in Germany will overcome their reluctance to invest any time soon given the huge uncertainty surrounding the direction of this country's long-term energy policy.[46]

Deutsche Bank's conclusion was stark: "The political class should therefore interpret private companies' reluctance to invest as a wake-up call."[47] It was a message Germany's political class didn't want to hear.

As the smoke cleared on the enormity of Germany's energy policy debacle, renewable-energy promoters and overseas regulators such as the EPA sought to blame the policy details to argue that they would not repeat the same mistakes in their countries. True, *Energiewende* constitutes a veritable catalogue of policy errors: promoting renewables in parallel with cap and trade; having an escalating scale of feed-in tariffs that overrewarded the least efficient technology (i.e., solar); not cutting subsidies when solar PV prices plunged thanks to Chinese imports; the 2009 EEG amendment that tapered down subsidies, sparking a stampede to lock in high tariffs with obsolescing technologies; rewarding poorly located wind farms with higher tariffs; incentives for self-consumption by wealthy households so the less well-off are burdened with a higher proportion of grid costs. In principle, wise policymakers not in hock to green interests would avoid all these mistakes. That would miss the big picture. The Greens and their

Far Left allies in the SPD wanted renewable energy to bring about the deindustrialization of Germany, a vision articulated by Wilhelm Ostwald in the first decade of the last century and subsequently revived by Hermann Scheer. In that sense, *Energiewende* is working to plan.

For those who don't share this vision, there are problems inherent in renewables technology that perfect policy execution and the wisdom of a King Solomon could not overcome. Even if the cost of wind and solar fell so they no longer needed direct subsidy (so-called grid parity), adding wind and solar PV capacity would still make electricity more expensive. Integrating large quantities of intermittent, weather-generated electricity requires more grid infrastructure, poses major operational challenges, and destroys incentives to invest in the capacity needed to assure continuity of supply.

Grid infrastructure is not cheap. In Britain, the cost of grid reinforcements to integrate wind and solar are of the order of £8 billion ($12.6b), implying a doubling of the cost of National Grid's transmission network. Connecting and integrating electricity from offshore wind farms is reckoned to cost a further £15 billion ($23.7b). Altogether, this implies a near tripling of grid infrastructure costs.[48] Texas, America's leading wind state, experienced grid congestion caused by transmitting electricity from wind farms in the rural north and west of the state to Dallas and Houston. To help out wind-farm investors, the state had consumers fund nearly 3,600 miles of transmission lines under the inaptly named Competitive Renewable Energy Zones (CREZ) grid program. Subsidizing renewables via grid infrastructure spending can be more costly than overt plant-level subsidies. Bill Peacock and Josiah Neeley of the Texas Public Policy Foundation reckon that CREZ costs attributable to wind amount to $6.55 billion, compared to $4.14 billion for ten years of plant-level subsidies.[49]

There is a still more fundamental factor and that is to do with the physics of electricity. Electricity is unique. Unlike any other commodity, it has to be produced the instant it is consumed and consumed the instant it is produced. Since Benjamin Franklin's time, it has been known that thunderclouds store electricity. It is the vast scale of a thunderstorm that makes possible storage of an appreciable electrostatic charge. Yet the amount of energy involved in a single lightning flash might come as a surprise—comparable to that used to keep five 100-watt light bulbs alight for one month but concentrated into a mere 300 milliseconds or less.[50] The explosive nature of a lightning flash gives a clue to the problem of storing electricity: If you try to stuff a force into a small space (i.e., store it), the more tension it creates, the greater the costs, and the higher the danger. In due course, further stuffing becomes a physical impossibility: There is a sudden explosive release of energy. The channel temperature in a lightning flash reaches well over 9,000°C (16,200°F). You can't make a bucket and fill it up with electrons; if you did, it would be a bomb, not a battery.

Storing electricity therefore requires converting it into other forms of energy: as chemical energy in batteries or as potential energy in pumped-storage hydro systems, to be reconverted the moment it is needed, involving energy losses on the way in and the way out. By definition, this is inefficient and expensive. Countries can't function safely and efficiently without buffer stocks of commodities. According to energy and tech expert Mark Mills, at any given time, country-level supply chains of critical commodities typically have three months' worth of annual demand in storage. The annual output of Elon Musk's planned $5 billion "gigafactory" in Nevada, slated to produce more than all the world's existing lithium batteries combined, could store about *five minutes* of annual U.S. electricity demand. "Storing electricity in expensive short-lived batteries is not a little more expensive but tens of thousands of times more expensive than storing gas in tanks or coal in piles adjacent to idle but readily available long-lived power plants," Mills explains.[51]

Lack of storability makes the operating and economic dynamics of electricity generation and distribution entirely different from other forms of energy such as oil and gas, and from all other commodities: Supply must respond almost instantaneously to changes in demand; not enough, and there is a danger of degraded quality and power cuts; too much, and the transmission system can be damaged, wires deformed or even melted. Failing to equalize demand and supply can also lead to changes in frequency—too high, and it can damage appliances; too low, and equipment can underperform. It can also lead to the alternating current system getting "out of phase," which can lead to catastrophic outcomes, including exploding transmission system transformers.

The interconnectedness of the grid means that everything on it affects everything else. Weather-dependent wind and solar capacity transfers weather risk to power stations, at the same time taking revenue from them—in a sense, reversing the logic of the Industrial Revolution. Wind and solar power's intermittency, unpredictability, and variability mean regressing from industrial production, where, like a factory, outputs are precisely controlled by varying the inputs, to arable farming, where output is heavily dependent on the vagaries of the weather. This is the least well understood but most damaging consequence of renewable energy. It threatens to disrupt the defining technological accomplishment of the twentieth century.

14

The Curse of Intermittency

> As we look at engineering breakthroughs selected by the National Academy of Engineering, we can see that if any one of them were removed, our world would be a very different—and much less hospitable place.
>
> *Neil Armstrong, 2000*[1]

> There is no other choice: government either abstains from limited interference with the market forces, or it assumes total control over production and distribution. Either capitalism or socialism; there is no middle of the road.
>
> *Ludwig von Mises*[2]

At the National Press Club in Washington, D.C., on February 22, 2000, the first man to ever walk on the moon declared, "I am and ever will be a white socks, pocket protector, nerdy engineer born under the Second Law of Thermodynamics."[3] "Science is about what is. Engineering is about what can be," Neil Armstrong told the gathering organized around the National Academy of Engineering's top 20 engineering achievements of the twentieth century. The automobile took second place and the airplane third (the innovations spawned by space exploration twelfth). The internet (13), air conditioning and refrigeration (10), television and radio (6), and electronics all rely on the greatest achievement of all—"the vast networks of electricity that power the developed world."[4] According to Armstrong, the maturity of the other 19 achievements would not have been possible without electricity. "If anything shines as an example of how engineering changed the world during the twentieth century, it is clearly the power we use in our homes and businesses."[5]

Harnessing the energy that nature had stored in clouds and making it ubiquitous is an astonishing achievement. Maintaining voltage and frequency within narrow parameters is a technological feat in itself. The first computers were used to help manage electricity flowing into and around the grid, as it requires tight coordination to ensure exactly the right amount of electricity is being generated.

The economics are uniquely complex too. Different generators have different costs, reflecting differing costs structures (the ratio of fixed to variable costs) and different primary fuel sources—coal, gas, oil, uranium, and so on. Thus, different types of generating technologies have different optimal uses in meeting different levels of demand through the day, from baseload to peak. Add in location, and an integrated electricity system has an enormously complex production function to ensure that the optimal generating mix is being used. This gives rise to what Lion Hirth, one of the most acute analysts of the electricity industry, calls the paradox of electricity as an economic good. On the one hand, electricity is perfectly *homogeneous*. Consumers have no way of distinguishing electricity from different sources. Electrical current from one generator supplied into an electricity pool is perfectly substitutable for current from another at that point in time. But electricity is *heterogeneous* over time. The moment demand changes, the optimal generating mix changes and the wholesale price changes.[6]

This heterogeneity is reflected in how wholesale electricity markets work. European power exchanges typically clear markets each hour in each bidding zone. U.S. markets often clear the market in steps of five minutes in each node of the transmission grid. This makes the electric wholesale market unlike any other. There are 100,000 prices a year in Germany and 3 billion in Texas.[7] Except to a very limited extent, grid-scale electricity cannot be stored, so the possibility of price arbitrage is extremely limited. The result is extreme price variability compared to all other commodities. In Germany in 2012, for example, wholesale prices varied by a factor of two within a normal day.[8] By contrast, from its low in September 1986, the U.S. Gulf Coast Conventional Gasoline spot price took 2.5 years to double in April 1989 and nearly 16 years (to March 2005) to double again. The most intense period of price volatility in the U.S. Gulf Coast Conventional Gasoline spot price was during the post-Lehman financial crisis, when investors sought refuge in real assets, and the price doubled in 14 days.[9] Even this panic-induced volatility is slight compared to what is experienced in the wholesale electricity market as a matter of course, where there can be a difference of up to four orders of magnitude (1 to 10,000) between the high and low hourly prices in a typical year.[10]

Wholesale prices drive both short-term scheduling decisions on which generators should supply electricity in response to changes in demand (the merit order curve) and long-term investment decisions. "Heterogeneity of electricity has not only shaped market design but also technology development," Hirth observes.

> For homogenous goods, one single production technology is typically efficient. In electricity generation, there is a set of generation technologies that are efficient. "Base load" plants have high investment, but low variable costs; this is reversed for "peak load" plants. The latter are specialized in only delivering electricity at high prices, which rarely occurs. If electricity was a homogeneous good, no such technology differentiation would have emerged.[11]

For this reason, the most efficient way of supplying electricity to the grid is with a mix of technologies that varies with demand and responds to changes in the relative prices of primary energy sources. Setting prices for all generators at the price charged by the generator dispatching electricity at the highest point in the merit order should enable generators with lower variable costs further down the merit curve to recover their fixed costs and remunerate investors for their capital.

Modern life depends on a stable grid providing always available electrical power at the flick of a switch, and society puts a very high value on having a reliable grid. Unlike coal, oil, gas, and nuclear, the output from wind and solar can't be varied with demand.* They can't be counted on for rapid "dispatch" to maintain grid stability but make the grid less stable. Evaluations that ignore or downplay the intermittency penalty of wind and solar systematically understate how much they cost compared to conventional dispatchable capacity. Compar-

* When the wind is blowing, in principle, wind output can be reduced, so it could be said to be responsive to lower demand—but only when the wind is blowing. In practice, wind is not deployed to provide negative balancing capacity.

isons of the whole-life, so-called levelized costs of electricity of dispatchable plants, such as coal, oil, gas, nuclear power stations, and certain forms of hydropower with nondispatchable wind and solar, are, in the words of MIT energy economist Paul Joskow, "seriously flawed."[12] As Joskow explains in a 2010 paper, a MWh of electricity supplied in a period of peak demand can be worth more than one supplied in the middle of the night even though it cost more to produce. "It should be clear," Joskow concludes, "that using traditional levelized cost calculations to compare dispatchable and intermittent generating technologies or compare different intermittent technologies is a meaningless exercise and can lead to inaccurate valuations of alternative generating technologies."[13] The promotion of wind and solar technologies on the basis of comparison of levelized costs and, in the case of solar, the mirage of grid parity—when solar's levelized costs will purportedly match those of coal and gas—is no more than the sales patter of the huckster and the charlatan.

It comes as little surprise that in a July 2015 article, green billionaire and NextGen cofounder Tom Steyer claimed that renewables were competing head-to-head with fossil fuels and winning. "On a level playing field, these renewable resources are already in many cases the cheapest electricity options for new installations," Steyer wrote.[14] A level playing field would mean incorporating the astronomical costs of storing wind and solar electricity or, as a practical alternative, the full costs of idling coal and gas-fired power stations for when the weather changes and wind and solar aren't producing enough electricity. Similarly, Apple's claim that data centers run on 100 percent renewable energy remains untruthful until it makes a genuine scientific breakthrough and finds a way of storing gigacoulombs of electric charge as easily as it stores gigabytes of data.[15]

By contrast, Google engineers publicly acknowledged the impossibility of making renewables competitive with coal. "Unfortunately, most of today's clean generation sources can't provide power that is both distributed and dispatchable," they reported. Solar panels could be put on every rooftop but can't provide power if the sun isn't shining. Even assuming "the most extraordinary success possible," they came to the "frankly shocking" realization that such a breakthrough would not be enough to solve the climate problem.[16] Meanwhile Microsoft founder Bill Gates had reached the same conclusion. "Power is about reliability. We need to get something that works reliably," Gates told the *Financial Times* in June 2015.

> There's no battery technology that's even close to allowing us to take all of our energy from renewables and be able to use battery storage in order to deal not only with the 24-hour cycle but also with long periods of time where it's cloudy and you don't have sun or you don't have wind.[17]

What might now seem plainly obvious wasn't to early backers of intermittent renewables. In a 1991 paper, Cambridge University's Michael Grubb acknowledged that "renewable sources would be virtually ruled out if extensive storage really were an essential component."[18] Grubb demonstrated mathematically that large amounts of renewable capacity could be integrated without significant penalty. Subsequent experience refuted Grubbs's math. Twenty years later, the CEO of National Grid was saying that people in Britain would have to get used to having power only when it is available. "We are going to change our behaviour and consume it when it is available and available cheaply," National Grid's Steve Holliday said, anticipating a Great Leap Backward with diesel generators in every back yard to keep the lights on.[19]

Hermann Scheer was closer to the mark than the Cambridge professor. While biomass energy carriers could substitute a large share of fossil energy resources, a basic disadvantage of other "solar energy carriers" (i.e., wind and solar) was that they did not produce any fuels directly. Scheer's solution was similar to the one Hermann Honnef and Franz Lawaczeck pressed the Nazis to adopt in the 1930s. Solar hydrogen production was a must-have, Scheer wrote in *A Solar Manifesto*. There would be periods when there is "the risk of downtimes" or when not enough solar energy was being produced. Hydrogen could then be used as "stored sun." Especially in summer, there would be other times when too much is being produced.[20] Scheer admitted there was a difficulty. "The introduction of solar hydrogen is faced with the particular problem that the technologies for both hydrogen conversion and hydrogen utilization still need to mature to the volume production stage."[21]

Without viable hydrogen technology to siphon off excess wind and solar power, the influx of random amounts of heavily subsidized, zero-marginal-cost electricity onto the grid plays havoc with the wholesale market. Wind and solar have high capital costs but negligible variable costs because their energy inputs are free. In favorable weather conditions, they push dispatchable generators up the merit curve because they pay for their fuel. Feed-in tariffs mean wind and solar investors receive a guaranteed price and are insulated from the effect of their output on depressing the wholesale price. In the United States, wind investors hoover up a $23 production tax credit per MWh of electricity, so they can make money even if the wholesale price falls below zero. During periods of low demand, Denmark, Canada, and California as well Germany have experienced periods when excess renewable outputs push wholesale prices to less than zero. As with garbage, there is a collection cost, and battery-stored electricity is still more expensive than firing up a gas turbine. "These distortions are prognosticated to become even more pronounced in the future," the OECD warned in 2012.[22] Insulated from the pricing effect of their output, the outcome is massive overinvestment in renewables and underinvestment in the dispatchable capacity needed to keep the grid stable and the lights on.

Visible subsidies are the tip of the wind and solar iceberg. In his 200-page 2014 thesis on the economics of wind and solar, Lion Hirth was the first to rigorously model the worsening economics of wind and solar as they scale up. At 30 percent penetration, the curse of intermittency means the value of electricity from wind farms is worth only half that if wind were stable and nonintermittent, and the wholesale price of electricity—one received by coal and gas-fired power stations —falls to zero for 1,000 hours a year, representing 28 percent of the electricity produced by wind farms.[23] Solar's value decline is even steeper because it is more peaky. Compared to a steady source, solar PV's value is halved before it achieves a 15 percent market share.[24] These results are not a function of market design, which could be rectified by a few tweaks here and there, Hirth notes. Rather they are a direct consequence of the inescapable characteristics of wind and solar generation.[25] In the real world, Hirth finds that wind and solar generators produce disproportionately more power in regions of low electricity prices and produce disproportionately more power at times of low electricity prices.[26] At the then-prevailing €68 ($76) per MWh cost of windpower, Hirth estimates wind's optimal share of the

Northwest European market to be around 2 percent.[27] Solar is a lot worse. Even if solar costs fell 60 percent below their prevailing level, the optimal solar share would still be close to zero.[28] It's a long stretch from the optimal renewable market share of zero to 2 percent and the 20 percent envisaged by the EU's renewables targets.

While the costs of weather forecast errors and grid-related costs matter, Hirth concludes that

> the most important economic consequence of wind and solar variability is reduced utilization of the capital embodied in the residual power system, mainly in thermal plants.[29]

Hirth's analysis helps explain the value destruction inflicted on the three quoted German utilities mentioned in the previous chapter. As more wind and solar capacity is connected to the grid, investors in conventional generating assets are exposed to lower and more volatile wholesale prices. Power stations run for less time, produce less electricity, and receive lower prices for the electricity they dispatch, but the intermittency of renewables mean they have to cycle (ramp up and down) more often, incur higher maintenance costs, and have shorter operating lives. Lower revenues, higher costs, and greater uncertainty —a lethal combination that can be expected to set off a similar process of *destructive* destruction wherever large amounts of subsidized zero-marginal-cost intermittent capacity are put on the grid.

Investors who saw their capital destroyed in gas- and coal-fired power stations are hardly going to be lining up to put fresh capital into a new generation of efficient gas-fired power stations. The distortions caused by subsidizing renewables raise the question as to whether the wholesale market becomes functionally worthless in signaling investment in the new power stations that keep the lights on. This is a concern raised by the OECD in its 2012 report, asking whether the wholesale market would remain "the relevant instrument for matching supply to demand and for co-ordinating investment decisions."[30] The growing wedge between falling wholesale prices and the income received by wind and solar investors puts a question mark over "the very role of the marketplace to provide adequate signals for power generators," the OECD cautioned.[31]

How to keep the cities, homes, and data centers lit when there isn't enough sun and wind? Governments and utility regulators will be increasingly forced to intervene to make it profitable to invest in dispatchable capacity to rectify the distortion created by subsidizing renewables, driving costs still higher. Over eight decades ago, the Austrian economist Ludwig von Mises warned of the spiral of intervention—the first begetting unintended and unwanted consequences bringing forth another and another, the final unintended consequence being full state control. The logic of central planning of electricity leads to the extinction of the market. The role of wholesale prices as investment signals and the market as a discovery mechanism is taken over by regulators. Because what happens on one part of the grid affects everyone else, the unintended consequences of one intervention to rig the market begets another one, then another to address the unanticipated consequences of the second. This turns all providers of investment capital into rent seekers and lobbyists, the value of their investments dependent on the shifting sands of government policy. As we shall see in the next chapter, the hollowing out of the market to the benefit of wind and solar at the expense of the rest provides the rationale for the existence and feeds the growth of the Climate Industrial Complex.

The web of regulations and interventions becomes so complex that no one can understand how they interact or the consequences of each for the system as a whole. Commenting on the state of energy policy in Britain, leading energy economist Dieter Helm observed in 2015 that it was "quite hard to make it worse than it currently is." The situation had become so bad that "the CEGB [the nationalized Central Electricity Generating Board, founded in 1957] would actually be better than what we've currently got."[32] The logical endpoint of interventions to support renewables is state ownership and state control.

And what of cutting carbon dioxide emissions, the ostensible reason why governments intervened in the first place? A 2014 Brookings Institution report by Charles Frank compared the costs and benefits of decarbonization for wind, solar, and combined-cycle gas turbine (CCGT) plant on the basis of a $50 per metric ton cost of carbon. The analysis, which did not explicitly incorporate the extra grid and scale costs of renewables, found that wind generated annual net disbenefits of $25,333 per MW of capacity and solar generated $188,820 of annual net disbenefits, whereas one MW of CCGT capacity generated net benefits of $535,382 a year.[33]

In the light of Frank's analysis, a rational approach to reducing power sector greenhouse gas emissions would have encouraged expansion of CCGT capacity. At the beginning of the decade, Siemens supplied two of the most advanced CCGTs in the world to the Irsching power station in Bavaria. "The transition to a new energy policy is the project of the century for the Germans," the chief executive of Siemens's energy division said in March 2012. "But in terms of implementation, most of the work still lies ahead of us."[34] It certainly did. Three years later, the owners of the Irsching 4 and 5 plants gave notice that they would be closing both. Costing around a billion euros with a combined capacity of 1,400 MW, Irsching 5 had started operating only in 2010. Its 59.7 percent fuel efficiency made it one of Europe's most efficient, and Irsching 4, which became operational a year later with 60.4 percent efficiency, was one of the most efficient in the world. Neither had supplied any merchant power at all in 2014, and their output had been dispatched only under contract with the network operator to stabilize the network in response to temporary fluctuations. After expiry of this contract, the two CCGTs would have to cover all their costs by supplying merchant power. "Low wholesale price and rising renewables feed-in, however, render this impossible," the plants' owners stated.[35]

Visiting a solar company in April 2014, an exasperated SPD leader and vice chancellor, Sigmar Gabriel, exclaimed that the complexity of *Energiewende* had been completely underestimated. "Other European countries think we're nuts," the vice chancellor told his astonished hosts.[36] Closure of state-of-the-art CCGTs might not have been the intention—if it had been, Irsching's owners would not have spent a billion euros on them—but it was a direct consequence. Asked what message he would give other countries thinking about whether to adopt renewables, Fritz Vahrenholt offers blunt advice: "Don't follow Germany into this dead-end."[37]

A decision tree based on German experience shows the difficulty for other countries to find a rational basis to justify large-scale renewable programs (Table 2). Instead they opted to join the Gadarene swine. Others saw it differently. To *New York Times* columnist Thomas Friedman, Germany deserved the Nobel Peace Prize for setting off the stampede. During a visit to Berlin in May 2015, Friedman called *Energiewende* "an undiluted success." He lauded Germany for creating

Table 2: The Renewables Decision Tree—The hoops you must jump through to have a rational basis for committing to renewables above 10%

Question	Answer	Outcome
1. Is having renewables more important than cutting CO_2?	Yes	**Go to 2.**
	No	Renewables are highly inefficient at cutting CO_2 emissions, if they cut them at all. Under a cap-and-trade regime, their biggest effect is to displace CO_2 emissions, not cut them. **Renewables policies are not for you.**
2. Do you want renewable capacity to be more than 10% of generating capacity?	Yes	**Go to 3a and b.**
	No	Conventional electricity systems networks are robust and capable of integrating renewables, but only if renewables make up less than 10%.
3a. Are you happy for consumers to pay a lot more for electricity than they need to?	Yes	**Go to 4.**
	No	Germany's first Green energy minister said renewables would cost the equivalent of scoop of ice cream a month—turns out it was a €300 ice cream. **Renewables policies are not for you.**
3b. Do you want your energy intensive industries and manufacturing businesses to relocate overseas?	Yes	**Go to 4.**
	No	German manufacturers are making a slow, silent escape to maintain their global competitiveness. **Go back to 2.**
4. Do you want to keep the lights on?	Yes	**Go to 5.**
	No	**Go to 6.**
5. Do you have abundant, nonseasonal hydroelectric capacity?	Yes	You must be Norway, with a small population and plentiful year-round rainfall. **If not, answer the question again!**
	No	**Go to 6.**
6. Do you have friendly neighbors you can pay to take away your surplus wind and solar energy?	Yes	You'd better hope they keep off renewables and avoid having the same problem as you. **Go to 7.**
	No	**Go to 8.**

<table>
<tr><td rowspan="2">7. Do you have friendly neighbors who will guarantee to sell you peak power whenever you need it?</td><td>Yes</td><td>You'd better hope they keep off renewables and not have the same problem as you.
Go to 9.</td></tr>
<tr><td>No</td><td>**Go to 8.**</td></tr>
<tr><td rowspan="2">8. Do you believe in the "hydrogen society"?</td><td>Yes</td><td>So did Franz Lawaczeck in the Nazi era—but we are no closer to solving the electrochemistry of hydrogen production, and the storage and transportation of the smallest molecule in the universe.
Answer the question again.</td></tr>
<tr><td>No</td><td>**Go to 9.**</td></tr>
<tr><td rowspan="2">9. Do you have a working "smart grid" controlling the electricity consumption of billions of devices and appliances in millions of homes and businesses?</td><td>Yes</td><td>You don't! Forget the baloney about the Internet of Things. Controling gigawatts is not the same as controling gigabytes. As yet, no systems exist at any price to do that—and if everyone responds the same way to a price spike, the grid collapses.
Answer the question again.</td></tr>
<tr><td>No</td><td>You're honest—**go to 10.**</td></tr>
<tr><td rowspan="2">10. Do you have millions of electric-powered vehicles connected to the grid to store and discharge electricity when required?</td><td>Yes</td><td>You don't! Tesla hype aside, batteries make terribly expensive and limited ways for nearly all forms of transportation.
Angela Merkel set a goal of one million electric vehicles on German roads by 2020. In 2013, only 7,114 were sold.*
New science is needed just as much as Google and Bill Gates said it is for the grid. **Answer the question again.**</td></tr>
<tr><td>No</td><td>**Go to 11.**</td></tr>
<tr><td rowspan="2">11. Are you willing to pay the owners of coal- and gas-fired power stations billions of dollars/euros/pounds to produce very little electricity and keep the generators idle for when the wind drops or it clouds up?</td><td>Yes</td><td>**Go to 12.**</td></tr>
<tr><td>No</td><td>**Go back to 2.**</td></tr>
<tr><td colspan="3">12. Congratulations! You are renewables ready!</td></tr>
</table>

* Alexander Wendt, *Der Grüne Blackout; Warum die Energiewende Nicht Functionieren Kann* (The Green Blackout: Why the Energiewende Cannot Work) (Munich, 2014), p. 29.

China's solar panel industry, hailing it as a "world-saving achievement," though Friedman did not mention that German power station emissions had been rising.[38]

Most telling is what Friedman's article says about the destination of the journey taken by the Sixty-eighters. In Paul Berman's telling, there is something idealistic, courageous, heroic even, in the path some of them took.

> Millions of people had gone through the left-wing experiences of the sixties, and a distinct cluster of the most prominent and irrepressible of those people had drawn some very similar lessons and had travelled more or less the same path, from 1968 to NATO and the Kosovo War.[39]

By the summer of 2003, the humanitarian interventionism some of them espoused was a short distance from the position George W. Bush was to set out in a speech to the National Endowment for Democracy in November that year. Others, Berman says, were more cautious and respectful of international law. The prospect of bringing the two groups together was destroyed by a suicide truck bomb that blew up the UN's headquarters in Baghdad that August. The Sixty-eighters' path that engaged Berman's attention ended in the rubble of Baghdad's Canal Hotel.

There was another path running through the New Left's transformation into the Greens. "I have a friend who comes home," Oliver Krischer, vice chairman of the parliamentary Greens, told Friedman, "and, if the sun is shining, he doesn't even say hello to his wife. He first goes downstairs and looks at the meter to see what [electricity] he had produced himself."[40] This is where journey from the student protests of the 1960s had led: Returning from the office and greeting the *hausfrau*—so that's what happened to the teachings of Adorno, Marcuse, and the Frankfurt School!—is subordinated to seeking absolution from the symbol of a guilt-free lifestyle, a secular religiosity that, unlike the crucifix of German Catholic households, has the added benefit of producing a bit of income to salve the conscience of being well-off in one of the world's wealthiest societies. A solar panel on the roof makes materialism moral. At the same time, it is deeply subversive. It deploys values conducive to capitalism—the ethic of thrift and counting the pennies combined with a spurious sense of self-reliance—to hollow out capitalism from the inside. Feed-in tariffs transfer wealth from poor to rich, from those who don't have roofs to those who do; they foment an antibusiness culture based on phoney grid independence; they align the interests of vocal consumers with those of crony capitalists and green rent seekers and promote green consciousness among a class of white-collar professionals working in or for the public sector, hostile to the interests of German industry that generates the country's wealth. Hermann Scheer and his fellow Renewable Putschists had set in motion a genuine revolution. First Germany, then Europe, now spreading across the world.

15

Climate Industrial Complex

> It has not been appreciated until the past few years that the population of capillary endothelial cells within a neoplasm may constitute a highly integrated ecosystem.
>
> *Dr. Judah Folkman, 1971*[1]

Angiogenesis: the production and growth of new blood vessels. The term is found in research papers in the first decades of the twentieth century and was revived in 1966 by Judah Folkman, a surgeon at Boston's Mass General Hospital. During cancer operations, Folkman noticed that large tumors always had plenty of blood vessels, whereas tumors lacking blood vessels tended to be very small and white and could lie dormant if they hadn't induced blood-vessel growth. Vascularization—the forming of new blood vessels to support tissues—seemed to make all the difference in the tumor's world: "A tumor without blood vessels was stuck—not dead, but dormant."[2] Folkman began to suspect that the vascular system was not just a plumbing system but a complex and very active organ.[3] It took nearly four decades for Folkman's intuition that cancer cells emitted something triggering vascularization to gain wide acceptance.*

Wind and solar PV display many similarities to tumors in living organisms. They contribute nothing of value. The larger they get, the

* Angiogenesis is an example of how settled scientific orthodoxy shuts down conflicting or novel ideas. According to Folkman's biographer, Robert Cooke, "skeptics were ensconced on advisory boards of grant-making agencies and on the anonymous peer review panels of scientific journals. They found it easy to snipe at Folkman's contentions about angiogenesis and cancer, often repeating the damning, dismissive statement that became a mantra: 'Your conclusions are not supported by your data.'" Robert Cooke, *Dr. Folkman's War: Angiogenesis and the Struggle to Defeat Cancer* (New York, 2001), p. 86.

more they subtract and the more aggressive they become, ultimately threatening the viability of the system they live off. Without a rich supply of blood, they would, like the tiny, white blobs noted by Folkman, remain small and harmless. Promoting angiogenesis is the function of the Climate Industrial Complex.

The nexus of the Climate Industrial Complex is the state, for the sheer scale of resources required to feed the growth of the renewables neoplasm can be delivered only through the state's powers to tax, regulate, legislate, and administer. These form, to borrow Walter Bagehot's distinction, the dignified functions of the Climate Industrial Complex. The efficient parts—that which gets the dignified part to do its bidding—constitute the climate Deep State. Dense networks connect state bureaucracies and regulatory bodies to universities, think tanks, NGOs, the media, special interest groups, financiers and their lobbyists, and religious institutions.

The strategy, tactics, and modus operandi of the Climate Industrial Complex were set out in a January 1986 speech by Günter Hartkopf, a long-serving state secretary in the German interior ministry. Speaking at a conference of fellow German civil servants, Hartkopf recounted a decisive meeting in 1975 between government representatives and senior industrial leaders. The industrialists had been objecting to burdensome environmental regulation, which, they said, was holding up investment. The civil servants had come armed with arguments culled from the Club of Rome's *Limits to Growth.* The industrialists had no answers. Bad arguments beat no arguments, and the industrialists never asked for another meeting.

Neutralizing industry was only a first step, Hartkopf observed. Environmental protection required lobbying from outside the federal bureaucracy, demanding more environmental regulations. Current and former civil servants occupied senior positions in environmental pressure groups, illustrating the interdependence between public administration and NGOs. This extended to the media. Journalists saw themselves as active supporters of environmental improvement, helping to raise the environmental consciousness of the public, something Hartkopf believed was essential. In addition to the reporting of so-called environmental scandals, the media also publicized articles in scientific journals. Most were written by civil servants, as professors and their colleagues are civil servants who happen to be working at universities, a feature of German life unchanged from the pre-1945 era. The sheer magnitude of articles and their technical or policy conclusions not only provided a valuable resource to the government bureaucracy; they swamped industry's ability to react, as they had no chance of responding with the same number and quality.

Hartkopf was disarmingly candid about the positive role of untruths and disinformation, or what he called "empty phrases" in pushing forward the environmental agenda. A widely used example was "ecological equilibrium," which Hartkopf said was meaningless. More sinister was disinformation to obscure the reality of policy trade-offs. The claim that ecology and the economy were not in conflict was one of those. "I myself have made this claim, knowing it to be less than truthful," Hartkopf admitted. Investing to protect the environment being good for companies was another. While both claims were untrue, they would continue to be with us, implicitly telling his fellow civil servants that they had his blessing to continue to mislead.[4]

At the time, the Greens were only six years old and the Sixty-eighters' march through the institutions had just begun. Twelve years later, when the first Red–Green coalition was being formed in Berlin, they were occupying key nodes of the Deep State. "In the universities, the student renegades of the late 1960s and 1970s were full of professors, department chairs, and even deans," Paul Hockenos writes in his history of the Berlin republic.

> The concerns of the social movements had been institutionalized in well-funded "peace research" institutes and environmental think tanks. Their specialists weren't "counter-experts" anymore, simply experts. Their research and analysis fed Greens policy in areas such as the energy sector, for example, where the Greens laid out detailed plans to switch over to renewable energies and meet—even exceed—the carbon dioxide emission benchmarks set by the Kyoto Conference on Climate Change.[5]

The Sixty-eighters didn't just march through the institutions. They created new ones. The first was the Öko-Institut of Freiburg in 1977, which emerged from the antinuclear movement. The Wuppertal Climate Institute was founded in 1991 by Ernst Ulrich von Weizsäcker, an SPD politician. A co-chairman of the Club of Rome, von Weizsäcker once told students to eat beetroot instead of strawberries to save the climate. The Red–Green coalition upped green institutional firepower in 2000 by creating Dena, the German energy agency, to promote renewable energy and energy efficiency. In addition to the federal government, other Dena shareholders include some of Germany's largest financial institutions, such as the KfW Group (a state development bank with a half-trillion euro balance sheet), Deutsche Bank, the German Cooperative Bank, and the Allianz insurance group. Dena became embroiled in a scandal at the end of 2014. It involved Stephan Kohler, its CEO since its founding and one of the highest paid of the green apparat, who earned €220,000 ($272,000) in 2013—more than Angela Merkel's salary. It led to Kohler's resignation.[6] American foundation dollars were also poured into German eco-institutions.[7] The Agora-Energiewende, founded in 2012, which provided Rainer Baake with a perch before his return to a key role in the federal bureaucracy, is financed by the Germany-based Mercator Foundation and the European Climate Foundation, which in turn is partly funded by the San Francisco–based ClimateWorks Foundation.

The most powerful is the Potsdam Institute for Climate Impact Research that fields some 50 climate professors. Founded in 1992 by Hans Joachim Schellnhuber, a leading participant at the Essen 1992 great transformation conference (Chapter 2), the Potsdam Institute is generously funded by the German government. As Merkel's top climate adviser, Schellnhuber developed a special line in climate alarmism. In 2009, Schellnhuber made what he called a below worst-case warming scenario leading to population collapse. "In a very cynical way, it's a triumph for science because at last we have stabilized something—namely the estimates for the carrying capacity of the planet, namely below one billion people," a Schellnhuber prediction that even the *New York Times* called apocalyptic.[8] Four years later, Schellnhuber was telling an Austrian newspaper of the possibility of an ocean heat belch. "That would shock-heat the first ten kilometers of our atmosphere—the layer that we live in and where weather occurs—by 36 degrees Celsius!"[9]

Perhaps sensing Schellnhuber's taste for the apocalyptic, in June 2015, Pope Francis appointed him to the Pontifical Academy of Sciences. At a Vatican press conference to launch the pope's climate encyclical, the atheist Schellnhuber invoked God's help to declare that clean energy was "available in abundance" before evoking a line of

Teutonic sun worship stretching back to Wilhelm Ostwald a century earlier.

> All we have to do is develop the means to properly harvest it and responsibly manage our consumption. . . . We are already blessed with one that works perfectly well and is free to all of us: the Sun. Photovoltaics, wind and energy from biomass are ultimately all powered by sunlight.[10]

Notwithstanding the paganism, *Laudato Sì* faithfully incorporates the climate teachings of the Potsdam Institute. There was an urgent need to develop policies substituting fossil fuels and developing sources of renewable energy. His Holiness did concede that adequate storage technologies still needed to be developed.[11]

Man-made climate change gave the Vatican a perfect opportunity to further its global redistribution agenda that had first been set out in the 1967 encyclical *Populorum progressio*. As we've already seen, Schellnhuber's economist colleague, the former Jesuit Ottmar Edenhofer and another of the leading lights at the Essen conference, has been frank about using climate as a tool for global wealth redistribution. "One has to free oneself from the illusion that international climate policy is environmental policy," Edenhofer told the *Neue Zürcher Zeitung* just ahead of the 2010 Cancún climate conference.[12] Edenhofer engaged in some redistribution of his own. Between 2007 and 2010, he received more than half a million euros from MISEREOR, the German Catholic bishops' organization (itself partly funded through the church tax), to support his climate research. The organs of the climate Deep State have long tentacles. MISEREOR's 2015 Lenten campaign aimed to raise awareness of climate change ("Think anew. Dare to change") in the run-up to the Paris climate conference.[13]

While Schellnhuber cultivates the image of an ascetic climate priest preaching renewable salvation from climate catastrophe, Germany's first green billionaire, the brash and colorful Frank Asbeck, unapologetically flaunts his wealth. Asbeck is famous for his three *schlosses* and his fast cars. "Everyone has an evil side," Asbeck told *The Oregonian* about his Maserati. "And someone has to use the last remaining oil in order for people to switch to solar."[14] Asbeck joined the Greens with Petra Kelly and Gert Bastian but quit active politics as a Green to make money being green, founding SolarWorld in 1988. Designing, manufacturing, and distributing solar PV panels, SolarWorld grew rapidly. In October 2007, its stock market valuation hit €5.36 billion. It was downhill from there. Profits peaked in 2008 and sales followed two years later, falling 66 percent by 2013. A flood of Chinese imports and too much debt saw the company rack up over a billion euros of losses in three years. Rather like Donald Trump, the secret of his success was not going bankrupt, he once told a journalist.[15] A February 2014 debt-for-equity swap, involving a €1.3 billion creditor haircut, left shareholders with five percent of the company. Although Asbeck kept his job and was bullish about prospects for expansion, around half the 20,000 solar jobs in Germany have disappeared since 2011.[16] By autumn 2015, the company stock was valued at €320 million, little more than four percent of its valuation eight years earlier.[17]

After the European Council's 2007 decision to impose renewables targets, it became pointless for experts and industry participants to question the costs and consequences of large-scale deployment of wind and solar. The policy had been decided and the facts were going to be made to fit the wind and solar fairy tale. It had the effect of making government bureaucrats and energy regulators more dependent on the Climate Industrial Complex for expertise and advice. Michael Grubb, the British academic who had mathematically demonstrated renewables would not need battery storage (a proposition that even the pope does not accept), exemplifies how the Climate Industrial Complex occupies the nodes connecting government, academia, and climate advocacy. At various—often overlapping—times, Grubb has been an IPCC lead author on climate policy (including the Fifth Assessment Report); senior adviser on renewable energy to Ofgem, the U.K. energy regulator; a member of the Committee on Climate Change tasked with enforcing the U.K.'s carbon dioxide budget under the Climate Change Act; a professor of energy and climate policy at University College London after a long stint at Cambridge; editor in chief of the journal *Climate Policy*; chair of Climate Strategies ("world-class, independent policy and economic research input to European and international climate policy"); chief economist of the Carbon Trust ("Our mission is to accelerate the move to a sustainable, low carbon economy . . . making the case for change to businesses, governments and civil society worldwide") and the Royal Institute of International Affairs's head of energy and environment at Chatham House.[18] Climate Strategies' selection of "collaborators and supporters" displayed on its website (funders are not separately identified) illustrates the breadth and interconnectedness of the climate networks sitting behind it. It includes three British government departments (the Foreign and Commonwealth Office, Department of Energy and Climate Change, and the Department for International Development) plus the Committee on Climate Change; two German federal ministries (Economic Affairs and Energy and Environment, Nature Conservation and Nuclear Safety); two supranational bodies (the European Commission and the World Bank); and a number of foundations including Mercator and the European Climate Foundation.[19]

And here a key part of the explanation for the single biggest government resource misallocation in history comes into view: a dense network of people defined by the same outlook, committed to the same goal, and, crucially, agreeing on the means of achieving it that operates as a monopoly provider of advice and expertise to governments. If the "renewables to avert planetary catastrophe" construct were sound, criticism would strengthen it. Questioning, testing, appraising alternatives—all threaten to undermine the cause that unites participants in their shared commitment, creating a culture in which skepticism is a capital offense and suspension of skepticism a requirement, as posing fundamental questions would cast doubt over the whole enterprise.

Its survival as a governing policy paradigm depends on the absence of challenge from its scientific substructure through to its policy-prescription superstructure: the inherently speculative nature of unverifiable claims about future catastrophe; the opaque and, at times, dishonest conduct and presentation of climate science; observed global temperatures that lag model predictions; the continued rise in man-made emissions despite a near quarter of a century of international climate negotiations; the reality that developing nations will carbonize before they decarbonize; the fundamental constraint imposed by the nonstorability of electric charge, which means fossil fuel power stations are required for the foreseeable future; and the increased expense of cutting carbon dioxide emissions with wind and solar over alternatives, which creates a set of unavoidable problems for which, as yet, there are no commercially viable solutions.

The echo chamber created by wall-to-wall climate propaganda is mirrored within government bureaucracies. The experts they commission and the advisors seconded to them are drawn from the same milieu, as are the authors of papers and editors of academic journals. Alternative ways of thinking are sidelined and pushed out of the mainstream, where they can be ignored. Thus, the policy papers put up to the political heads of government departments and the heads of regulatory agencies, who normally get their jobs because of their commitment to the cause, reinforce rather than challenge their misconception that renewable energy policies do anything other than make rent seekers rich.

In a 2009 paper "Survival of the Unfittest," Bent Flyvbjerg, professor of major infrastructure projects at Oxford University, asked why the worst infrastructure gets built. Flyvbjerg, the world's most cited scholar on megaproject planning, wrote that a characteristic of major infrastructure projects is that there is "lock in" or "capture" at an early stage, leaving analysis of alternatives weak or absent.[20] In reviewing possible explanations as to the benefit shortfalls and cost overruns of big infrastructure projects, Flyvbjerg found that political-economic explanations—project planners and promoters "deliberately and strategically overestimating benefits and under-estimating costs"—better fit the data than other possible explanations.[21] Project managers and planners "lie with numbers." Accurate forecasting is counterproductive. "The most effective planner," in the words of a 1989 paper cited by Flyvbjerg, "is sometimes the one who can cloak advocacy in the guise of scientific or technical rationality."[22]

Flyvbjerg comments that rich countries can afford infrastructure disasters but did not become rich by building them: "They do so when they have become rich."[23] Though worrying, the costs of the largest infrastructure projects examined by Flyvbjerg are small change compared to the colossal amounts being spent on trying to decarbonize industrialized economies with wind and solar. Typically the projects he examined are one to three orders of magnitude smaller (billions and tens of billions of dollars, pounds, or euros) than nation-scale renewable deployment (hundreds of billions up to trillions). Such colossal costs impose strains even on rich economies like Germany's.

What can be done? In the case of megaprojects, large forecasting inaccuracies have led to discussions about "firing the forecaster." Flyvbjerg goes further. "Some forecasts are so grossly misrepresented that we need to consider not only firing the forecasters but suing them, too—perhaps even having a few serve time."[24] Malpractice in project management, argues Flyvbjerg, should be taken as seriously as in other professions such as medicine and the law.

Government agencies and regulators are not in the business of making themselves accountable for their mistakes and their malpractice. When they make mistakes or fraudulent claims, the public pays. Picking up Flyvbjerg's insight, a partial remedy would be to inject incentives for honesty to redress the incentives for dishonesty identified by Flyvbjerg by requiring independent audit so that regulatory policies and their supporting analyses are subject to independent review. As with a securities filing, it should make explicit the reasonableness of the assumptions, uncertainties, and risk factors and have auditors legally stand behind their audit opinion.

Putting the uncertainties, risks, and unknowns into the public domain is all the more important given the forces trying to close down public debate. The effect of wall-to-wall government- and foundation-funded green propaganda is not so much to silence dissenting voices as to asphyxiate them. There is not a free debate in Germany, Vahrenholt says. It was the German government's delegate who had insisted on editing out the hiatus or pause in average global temperature from the IPCC's Fifth Assessment Report, on the grounds that it would confuse German voters.[25]

16

Power without Responsibility

> In combination with science, conservation includes many personal and subjective opinions. . . . Notably, the opinions of conservation scientists are sometimes in conflict with other environmentalists', because of different perceptions, priorities and scales of analysis.
>
> *Clive Hambler and Susan Canney*[1]

The big campaigning NGOs were latecomers to the global warming party. In his 1986 speech, Günter Hartkopf had highlighted the role of environmental NGOs. "A membership of about four million people can be mobilized whenever needed and is therefore an asset of great potential."[26] NGO mobilization is not driven by scientific or technical rationality but by fear and emotion. NGOs had played a limited but important role at the 1988 Toronto climate conference. According to environmental writer Fred Pearce, the World Resources Institute's chief lobbyist, Rafe Pomerance, had challenged fellow lobbyist and acid rain expert Michael Oppenheimer of the Environmental Defense League what they should do to set the international agenda. Oppenheimer, who had attended the second Villach conference the previous year, responded, "Offer a target." A 50 percent cut would have scientific credibility. Pomerance pushed back: "How about twenty percent by the end of the century and fifty percent eventually?" It was close to the target adopted by the conference.[27]

As late as 1989, Greenpeace International was being told by its scientists to stay away from the issue because the science was too uncertain. A Friends of the Earth activist told Pearce that, in campaigning terms, global warming was "very difficult to handle."[28] Greenpeace and Friends of the Earth had spent much of the 1980s campaigning against nuclear power. Safety concerns in the wake of Three Mile Island and Chernobyl and falling hydrocarbon prices had reduced the number of nuclear projects to a trickle. In Britain, preparations to privatize the electricity industry had revealed nuclear power's dire economics. By the end of the 1980s, Pearce recorded that many energy campaigners had been getting bored and looking for new challenges. They latched onto the greenhouse effect—"and were immediately faced with their old foes, the nuclear companies, desperate for orders, arguing that nuclear power would save us from the greenhouse horrors."[29] Global warming and the risk of letting nuclear return by the back door was a rite of passage for environmental NGOs. They had made their names with high-profile campaigns against nuclear testing, whale hunting, and seal clubbing—reducing complex disputes to a punchy headline and an iconic photograph. Instead of defining themselves by what they were against, they would also campaign for something. That something was renewable energy. The Climate Industrial Complex would have at its disposal the best shock troops money could buy.

Berlin, March 26, 1995. "Our gathering here today is a somewhat historic one," began Rolf Gerling, owner of a $1.6 billion reinsurance company he had inherited from his father. It was two days before the first conference of the parties to the UN Framework Convention on Climate Change, which would culminate in the Kyoto Protocol two and a half years later. "To my knowledge, this is the first time that representatives of the insurance, reinsurance, banking and pension-fund industries have gathered together to discuss the issue of climate change," Gerling continued. "That we do so in the company of representatives of the environmental groups, and the solar-energy industry, adds a further dimension of novelty—and perhaps risk—to the occasion."[2]

The meeting had been the brainchild of Dr. Jeremy Leggett, scientific director of Greenpeace. A geologist, Leggett had spent the early part of his career teaching petroleum geologists and petroleum engineers. "One summer I actually worked for an oil company."[3] After narrowly missing out on a top job at Friends of the Earth, in 1989 Leggett quit teaching how to find oil and joined Greenpeace to campaign to keep oil in the ground.

Greenpeace had always been the most militant of the direct-action environmental groups, cut from very different cloth from Washingtoninsider organizations such as the World Resources Institute and the Environmental Defense Fund. It had been formed in Vancouver from a group protesting against underground nuclear testing due to be carried out on an island at the end of the Aleutians in 1969. The tests could cause a chain reaction of earthquakes and tsunamis, Robert Hunter, one of the protest committee organizers, claimed—something Hunter later admitted he had never believed. "It's not that we ever lied," Hunter explained, "but we had painted a rather extravagant picture of the multiple dooms that would be unleashed." Not only tsunamis and earthquakes, but also radioactive death clouds, decimated fisheries, and deformed babies. "We never said that's what would happen, only that it could happen. . . . Children all over Canada were having nightmares about bombs."[4]

Fear was Greenpeace's metric of success, and manufacturing it was hardwired into Greenpeace from the beginning. In the mid-1980s, Greenpeace directors proposed a "ban chlorine worldwide" campaign, tagging the eleventh-most common element in the Earth's crust "the Devil's Element." Perhaps the other Greenpeace directors hadn't realized why seawater tastes salty. Patrick Moore, another cofounder and the only director of Greenpeace International with a formal education in science, argued that chlorine was the most important element for public health and medicine. Adding chlorine to drinking water had been the biggest advance in the history of public health. "This fell on deaf ears, and for me this was the final straw. I had to leave."[5]

During the 1980s, Greenpeace waged war against the Sellafield nuclear reprocessing plant in Cumbria. It blocked a discharge pipe, its boat being showered with court injunctions, but in 1989 it used the courts to extract a groveling apology from *Nature* and a large charitable donation after the journal had suggested it practiced ecological terrorism.* The same year, Greenpeace sold itself to the Kremlin when it opened its Moscow office, cofinanced by a state-owned record company. Under the patronage of a leading member of the Soviet Academy of Sciences, Greenpeace made it clear that it would have nothing

* Twenty-five years later, Greenpeace did commit an act of ecological vandalism when its activists entered a strictly prohibited area of the Nazca Lines in Peru and laid down a "Time for Change! The Future Is Renewable" banner beside the figure of a hummingbird as the UN climate talks began in Lima. Dan Collyns, "Greenpeace Apologises to People of Peru over Nazca Lines Stunt," www.theguardian.com, December 11, 2014, http://www.theguardian.com/environment/2014/dec/10/peru-press-charges-greenpeace-nazca-lines-stunt (accessed October 27, 2015).

to do with green groups in the Baltic republics. Recycling standard Soviet propaganda, Greenpeace denounced them as little more than separatist organizations.[6]

This, then, was the organization Leggett joined. One of his first tasks was getting 40 signatories, including two Nobel Prize winners, to protest against nuclear power as a solution to global warming. Leggett also persuaded Greenpeace to initiate its antiscientific campaign against genetically modified organisms (GMO), one that it has pursued for a quarter of a century, most recently with a groundless crusade against golden rice and its potential to have a dramatic effect on reducing childhood mortality caused by vitamin A deficiency in poor nations. "GE crops are prone to unexpected effects which can pose a risk to environmental and food safety," Greenpeace asserts in its anti–golden rice campaign, showing that its tactics still rely on Hunter's playbook of fear and dissimulation.[7]

Leggett realized Greenpeace needed to be more selective in its choice of enemies, change the perception that it was against all businesses, and find some business allies. He also saw that Greenpeace was way behind on global warming. The UN had set up the international negotiating committee for a framework convention on climate change in December 1990, which was signed at the Rio Earth summit in June 1992. Leggett was frustrated by Greenpeace's lack of impact. He spent a month in Brazil after the Earth summit mulling it over. Only big fossil fuel producers and consumers had been represented as the voice of business at climate events. Why wasn't the insurance industry there? "Their very solvency seemed to be at stake should the dice roll unkindly in an overheating world. I decided to refocus my efforts, post-Rio, on the financial services sector."[8]

It was a shrewd choice. Reinsurers are in effect wholesalers that enable insurance companies to diversify their risk exposure. They are big and well capitalized—in 2014, the top five reinsurers booked $118.57 billion of premiums—but are exposed to an underwriting cycle.[9] Excess capacity leads to softer premiums until a spike in natural disasters or other catastrophes shakes out capacity and drives up premiums. Talking up the risk of climate change disasters might help the most strongly capitalized reinsurers buck the cycle and firm up premiums. That was the theory. The reality was different. Over two decades later, Berkshire Hathaway, the world's fifth-largest reinsurer, was saying the effects of climate change, "if any," had not affected the insurance market. Owner Warren Buffett told CNBC in March 2014,

> it has no effect . . . [on] the prices we're charging this year versus five years ago. And I don't think it'll have an effect on what we're charging three years or five years from now.[10]

Nonetheless, Leggett won the propaganda war. Climate change was costing insurance companies "unbelievable" payouts, Secretary of State John Kerry said less than a year after Buffett had corrected the record. Indeed, they were unbelievable because they weren't being made. The lack of storm losses and the influx of new capital "has driven the cost of reinsurance for major categories of property and casualty losses far below what traditional underwriters have claimed are sustainable losses," John Dizard wrote in the *Financial Times*.[11]

To Leggett, climate disaster alarmism could have another effect beyond winning a strategic ally in the carbon wars. "In the heart of the Amazon rainforest, I daydreamed about the future flight of capital from carbon to solar, and schemed of ways to make that come about."[12] Leggett found a receptive audience when he gave the annual lecture of the Society of Fellows of the Chartered Insurance Institute in the City of London in November 1994. The battle to head off the threat of climate destabilization would be won or lost over the extent to which society ends the era of fossil fuel dependence and energy profligacy and replaces it with solar power and energy efficiency, Leggett told them.[13] He didn't have to "do a Hunter" and manufacture fear, as the Toronto climate conference had already "done a Greenpeace" six years before. All Leggett had to do was cite its claim that the impacts of a destabilized climate as being "second only to nuclear war."[14]

Where Leggett broke new ground was being the first to raise the threat of climate change to the insurance sector's solvency and the impact of decarbonization policies on the asset values of fossil fuel companies. "Global warming threatens the security of energy-sector investments in a way that is as yet little appreciated by investors." Investing in carbon-fuel industries was far from being a safe bet, Leggett warned.[15] Closing off the zero-carbon nuclear escape route ("nuclear power cannot compete economically with energy efficiency and most forms of renewables"), Leggett made his plug: "Saving the capital markets—not to mention the planet—will require kick-starting multi-billion dollar markets in solar energy."[16] Banks, insurers, and pension funds should send emissaries to the climate change negotiations. There were practical things they could do. Now that solar PV panels were cheaper than marble, the facades of their head offices could be covered in PV. "And it can all be made to look good in corporate advertising."[17] It was 1994, and Leggett had fired the first shot in the fossil fuel divestment campaign. "In a marriage between the manifest technical viability of solar power, and the redirection of capital flows by financial institutions, lie the seeds of a solar energy revolution."[18]

A year on from addressing Gerling's group, Leggett was giving the after-dinner speech of the congress banquet of the World Renewable Energy Congress in Denver. It was a busy time. He'd flown in from a European Union conference in Milan on setting a 15 percent renewable energy target by 2010. After Denver, it was on to California, then back to Michigan. Flying back and forth across America, Leggett pondered. He still couldn't envisage a specific set of circumstances forcing the river of "carbon-bound" capital to break its banks. "But I schemed—and I suppose dreamed—on." He had already decided he would leave Greenpeace. "I wanted to fight to divert the river of capital from inside the solar market." He didn't know what his new company would do but knew it would be called Solar Century.[19]

Founded in 1998, Solar Century had a slower start than Frank Asbeck's SolarWorld thanks to the delayed boom in ground-mounted solar plants in the United Kingdom, and in 2007 it had sales of only £13.9 million. A solar conference in London in June 2010 hyped Britain's imminent government-fed solar PV frenzy. "The solar industry and market in the UK is about to explode and increase over 100 percent for many years following the introduction of government incentives on solar energy in April," the conference organizers gushed.[20] In five years, the gusher of government support enabled Solar Century to increase its top line by an average 45 percent a year and more than double it in 2015 alone, when it racked up revenues of £222.3 million. Its growth was almost entirely fueled by government. As its 2015 annual report noted, Solar Century's biggest business risk was the possible reduction in government support for solar PV. It would try to redeploy resources to "grid-parity" markets in Africa and South America.[21] These markets accounted for 2.1 percent of Solar Century's

2015 revenue.[22] Thus, 97.9 percent of revenues and 100 percent of Solar Century's £10.7 million profit depended on government support in one form or another, making Solar Century a fine example of twenty-first-century green rent seeking. Not bad for the former Greenpeace activist who held 30.7 percent of Solar Century's shares.

That 2010 London conference brought together Jeremy Leggett and Hermann Scheer for the last time. Four months later, Scheer was dead. "I miss that hero," Leggett tweeted three years later.[23] Scheer had been Leggett's inspiration. "A giant of a man in terms of vision, energy and charisma," he had been the architect of the primacy of solar in Germany. With a poll showing 84 percent of Germans wanting to go 100 percent renewable as quickly as possible, Germany "is leading the global charge to renewables and we have him to thank for it," Leggett said.[24]

Other polls suggested that Germans were not as stupid or credulous as Leggett thought. Germans were already complaining about rising electricity bills, and Germany had become the world leader in the spread of fuel poverty. Between 2008 and 2011, the number of households trapped in fuel poverty (spending ten percent or more of their income on energy bills) rose by one-fourth.[25] "People here have to decide between spending money on an expensive energy-saving bulb or a hot meal," a representative of the Berlin office of the Catholic charity Caritas told *Der Spiegel* in 2013. More than 300,000 households a year were being disconnected because of unpaid bills. Energy efficiency and solar panels were for the better-off: "It is only gradually becoming apparent how the renewable energy subsidies redistribute money from the poor to the more affluent," the left-of-center *Der Spiegel* editorialized, pointing out that a person in a small rental apartment is subsidizing the solar panel on a homeowner's roof.[26] A 2011 Forsa Institute poll conducted in the wake of the Fukushima accident had only 39 percent of Germans wanting to go 100 percent renewable. Three years later, that figure fell to just 9 percent.[27]

Not supporting something is not the same as opposing it, and overt opposition to *Energiewende* is thin on the ground. The absence of opposition to green energy wasn't because Germans had lost their minds. It was because they'd lost their voice. According to Alexander Wendt, there is a sizeable minority of national and state politicians both in Merkel's Christian Democrats and in the SPD who are aware of *Energiewende*'s incredible cost and endless technical problems but dare not speak up and risk losing their seat in parliament. Only one party is openly against *Energiewende*, the Alternative for Germany (AfD), which narrowly missed the five percent hurdle in the 2013 elections. On the other side, Wendt reckons two-thirds of all journalists are supporters of the Greens or other parties of the left. "What we see in Germany is a perfect example of what the political scientist Elisabeth Noelle-Neumann called the spiral of silence: the more an opinion is under-represented in public, the more people tend to believe that this opinion must be weak and unacceptable," Wendt says, pointing to the most dangerous weapon in the armory of the Climate Industrial Complex, one which merits a chapter to itself (Chapter 23).[28]

NGOs were crucial to the Climate Industrial Complex's deployment of the spiral of silence. According to a 2003 study, *The 21st Century NGO* (project partners: the UN Global Compact and UNEP), globally the not-for-profit sector was worth over one trillion dollars.[29] Their power derives from their single greatest asset—a halo of public trust derived from the apparent purity of their motives. A 2002 survey of 400 American and 450 British, German, and French opinion leaders (high income, college educated, media attentive) conducted between January 2001 and June 2002 found that on specific issues such as human rights, health, and the environment, NGOs had a huge trust advantage over business, government, and the media.

European opinion leaders are considerably more credulous about NGOs than their American counterparts. NGOs took the top three places in Edelman's trusted brands ranking (Amnesty International, 76 percent; WWF, 67 percent; and Greenpeace, 62 percent), then a big drop to Microsoft in fourth place (46 percent).[30] American opinion

Table 3: Opinion leaders' trust ratings for institutions—U.S., U.K., Germany, and France (January 2002)

	Government	Corporations	Media	NGOs
Human Rights	14%	4%	14%	59%
Health	17%	7%	12%	54%
Environment	16%	6%	13%	55%

Source: Richard Edelman, "Rebuilding Public Trust through Accountability and Responsibility: Address to the Ethical Corporation Magazine Conference," New York, October 3, 2002, slide 9.

leaders ranked Microsoft first for brand trust (56 percent) with WWF, the top NGO, in seventh place (43 percent), Amnesty in tenth place (40 percent), followed by Greenpeace (38 percent).[31]

The logic of their respective rankings led governments and businesses to piggyback on green NGOs' trust luster. Thus the European Commission: It pays NGOs to lobby the commission and the European Parliament. The NGOs get cash—according to one study, the commission has paid more than £90 million in 15 years—and the EU's environmental agenda gets a big push. A commission spokesman justified this dubious arrangement, saying it was designed to ensure a "broad policy debate" (European use of English to indicate the opposite of the literal sense).[32] The largest green NGOs club together as the Green 10 to coordinate their European lobbying activities and promote EU environmental leadership globally.[33]

The Green 10 ran a joint manifesto for the 2014 elections to the European Parliament, pressing the EU to adopt ambitious and binding targets for—and note the order—"renewable energy, greenhouse gas emissions, and energy efficiency for 2030."[34] As we shall see, meeting EU energy efficiency targets led to one of the greatest government-induced public health disasters of recent times. Filings as of October 2015 show that the European commission paid the Green 10 a total of €5.9 million in lobbying funds for the most recent year, topped up with a further €942,214 from member states. Friends of the Earth Europe got the most (€1.69m from the commission plus €268,978 from member states). The next-largest commission grant, of €1.31 million, was to the CEE Bankwatch Network. Ostensibly monitoring the environmental impact of banks' financing activities in central and eastern Europe, CEE Bankwatch Network is well positioned to make life difficult for the governments of the Visegrad Four (Poland, Hungary, the Czech Republic, and Slovakia), which oppose the EU's climate and energy policies. WWF collected €657,014 in commission lobbying money.[35] (Greenpeace had a policy of not accepting government money except from the Soviet Union.)

Saving the planet also gives NGOs license that they deny to others. In June 2014, Greenpeace admitted that it had lost €3.8 million of donors' money in a currency bet that had gone wrong. Worse followed. Despite campaigning against the growth of air travel, which, it said, "is ruining our chances of stopping dangerous climate change," a week later it emerged that Greenpeace had been paying one of its executives for a twice-monthly air commute between Luxembourg and Amsterdam. "As for Pascal's air travel. Well it's a really tough one. Was it the right decision to allow him to use air travel to try to balance his job with the needs of his family for a while?" the organization's U.K. boss, John Sauven, blogged. "For me, it feels like it gets to the heart of a really big question. What kind of compromises do you make in your efforts to try to make the world a better place?"[36]

Were Greenpeace and the other environmental NGOs actually making the world a better place? Nature conservation was what the first nature organizations had been established to do. "All over the world today," reads the first sentence of the 1960 Morges Manifesto, which led to the establishment of the World Wildlife Fund, "vast numbers of fine and harmless wild creatures are losing their lives, or their homes, in an orgy of thoughtless and needless destruction."[37] Yet WWF and other NGOs became vocal advocates of wind farms that are causing a bat and avian holocaust. Because wind farms tend to be built on uplands where there are good thermals, they kill a disproportionate number of eagles and other raptors. According to Oxford University's Clive Hambler, the Tasmanian wedge-tailed eagle is threatened with global extinction by wind farms. In North America, wind farms are killing tens of thousands of raptors, including golden eagles and bald eagles.[38]

Few issues better exemplify the existential crisis of environmentalism than environmentalists' support for raptor-killing wind farms. Postwar environmentalism had come into being following publication of Rachel Carson's *Silent Spring* (1962). Environmentalism's cause célèbre was the worldwide banning of the pesticide DDT, which Carson and her followers claimed was poisoning birds ("The few birds seen anywhere were moribund; they trembled violently and could not fly. It was a spring without voices.")[39] Later studies suggested that the DDT metabolite known as DDE might cause thinning of raptor eggshells. However, no mechanism has been identified which shows how DDE might cause the thinning. The flimsiness of the alleged DDT link is apparent from a 1998 study that found eggshell thinning had begun 50 years before the introduction of DDT.[40] But when it comes to nonhypothetical causes of avian mortality from the blades of wind turbines, the green response is to say that cats kill more birds than wind farms.[41] A sophisticated variant is that fossil fuel and nuclear power stations will kill nearly 15 times as many birds per gigawatt hour than wind farms, an entirely bogus claim that found its way into the IPCC's Fifth Assessment Report.[42] Losses due to climate change are vastly more speculative than those due to observable current mortality and habitat loss. Climate change might benefit many birds in Britain, Hambler suggests.[43]

Wind farms kill bats when pressure waves from rotating blades cause their lungs to explode. One study cited by Hambler showed that bats killed by German turbines may have come from places 1,000 miles or more away, and some studies in the United States have put the death toll as high as 70 bats per installed megawatt per year. With 40,000 MW of turbines currently installed in the U.S. and Canada, this would give an annual death toll of up to 3 million bats. Such is the real-world impact of renewable energy, in contrast to the theorized effects of possible future warming. Most species threatened by "climate change" have already survived 10 to 20 ice ages and sea-level rises far more dramatic than any experienced in recent millennia or expected in the next few centuries, Hambler argues. "Climate change won't drive those species to extinction; well-meaning environmentalists might."[44]

A similar elevation of the hypothetical over the actual applies to forests. As we saw in Chapter 6, acid rain hysteria in Germany in the 1980s had been triggered by erroneous reports of the imminent death of German forests. The ecological destruction caused by renewable energy elicited barely a murmur of protest. As the best wind locations were populated with wind turbines first, turbines are increasingly being located in woods and forests. Each tower and its concrete foundation needs a cleared area of 6,000 to 10,000 square meters; roads need to be driven through forests (they must be straight to accommodate the 70-meter turbine blades), and the turbines must be sufficiently higher than the surrounding trees to be above the turbulence created by the wind blowing over the trees, resulting in pylons and blades standing 200 meters high.

Destruction of Europe's forests and woodland is also being accelerated by policies encouraging the use of biomass—timber, in everyday English—to displace fossil fuels. Under EU rules, fuelwood qualifies as zero-carbon on the grounds that the trees will grow back and carbon dioxide emitted from them will be reabsorbed. According to Hambler and coauthor Susan Canney, use of fuelwood is one of the most destructive of all energy sources. Increased use of bioenergy is leading to the destruction of "exceptionally important forest and other habitats," they write in their university-level textbook *Conservation*.[45] Businesses and households are incentivized to buy wood-burning boilers and rewarded for the amount of heat they generate. More wood is now being burned from British forests than at any time since the Industrial Revolution. According to Hambler, subsidizing biomass is "about the most bizarre thing you could do to counter climate change. . . . Big, hardwood trees are enormous carbon sinks and take hundreds of years to be replaced." Mark Fisher of Leeds University's Wildland Institute is more outspoken: Forests are "being butchered in the service of an ideology."[46]

Even so, Europe's forests are not large enough to feed the voracious appetite of the biomass subsidy seekers. With a nameplate capacity of 3,960 megawatts, the Drax power station, located in North Yorkshire, was once the largest coal-fired power station in Western Europe. Following part conversion to burning wood pellets, Drax's biomass is being sourced from American hardwood trees, such as oak, sweetgum, cypress, maple, and beech, 3,700 miles away in the forests of North Carolina. "We take the sawdust that's left over from sawmills that are cutting big trees that go into house building," Drax Power's chief executive claimed in a promotional video. Drax's claims have not gone unchallenged. A campaigner from the Dogwood Alliance fighting to conserve the forests of the South saw trees being cut all the way to the bottom before being loaded off to the pellet plant in Ahoskie, North Carolina. "Wildlife buzzed, chirped and splashed all around as huge hardwood cypress trees towered above—a testament to the incredible biodiversity that exists in the region," Adam Macon, who had tracked the timber back to where the trees had been felled, told Britain's *Mail on Sunday* in June 2015. "All that was left were the stumps of once great trees. They had destroyed an irreplaceable wetland treasure."[47] Whatever the impact on carbon dioxide emissions of switching from coal to biomass, the environmental impact on American woodland is unambiguously negative.

Not that destroying Southern woodlands rewarded Drax's shareholders. Anticoal policies had led Drax to switch from burning cheap imported American coal to expensive American biomass costing two and a half times as much. At the end of 2014, the stockmarket got wind of rumors of cuts in biomass subsidies. By July 2015, the Drax share price had more than halved—a fate that can befall any energy company that finds itself in the maw of the state—demonstrating that what matters is not how well investors make judgments on the efficiency of various technologies but how well they second-guess the twists and turns of government policy.

WWF exemplifies the NGO evolution from conservation to environmentalism, a trajectory also taken by the Sierra Club (founded in 1892) and the National Audubon Society (1905). The impulse for WWF's founding had come from fears that the decolonization of Africa and population growth would threaten African big game and its other wildlife. Hunting was the sport of princes. WWF had two, Britain's Prince Philip and Prince Bernhard of the Netherlands, and WWF saw hunting as a way of promoting game reserves. In 1968, WWF-Switzerland hired Roland Wiederkehr, a 25-year-old journalist who would work for WWF until 1987, when he became a Swiss MP. Influenced by the 1968 student protests, Wiederkehr began to campaign on environmental themes such as urbanization, pollution, waste, and consumerism. In 1973, WWF joined other Swiss environmental organizations to campaign against a proposed nuclear power station at Kaiseraugust near Basle. WWF and five other environmental organizations published a 200-page energy manifesto: "Our energy policy is a mirror of our relationship to the environment: ruthless exploitation or preservation of nature?"[48] In the 1970s, WWF-Switzerland was joined by other national WWF organizations embracing environmental activism.

WWF's transformation into an environmentalist NGO was decided in the early 1980s. "It would be much too simple and mere alibi just to raise funds for the panda and the elephant while lying over conservation issues at home by keeping polite silence on them. . . . We might just as well raise funds for dogs' and cats' homes," Daniel Hüssy, one of the WWF's founders, said in 1981 in defense of the Swiss WWF's antinuclear campaign.[49] Hüssy was supported by another cofounder, Max Nicholson, who argued at WWF's twentieth anniversary conference that while WWF had made vast and successful efforts to save tigers and rhinos, its conservation efforts would be futile unless it addressed what Nicholson called the "three Nasty Giants" undermining the future of life on Earth—technological development, energy, and population growth. "The sad truth is that someone will have to tackle those three big nasties, and if it isn't us, who will?"[50]

In January 1982, representatives of WWF-UK, WWF-Netherlands and WWF-Switzerland met at Hüssy's Zurich office and, in Hüssy's words, "decided that there had to be a revolution"; WWF must take up an uncompromising environmentalist position.[51] Later that year, WWF decided that "it was pointless to treat symptoms while ignoring the disease, and that the WWF must therefore address the root causes which threatened the survival of nature, and hence of man."[52] WWF's European branches had difficulty with the concept of wildlife, and environmentalism's triumph was sealed with its name change four years later, when it became the World Wide Fund for Nature.* In the light of WWF's active and well-funded support for environmentally destructive renewable energy, World Wide Fund *against* Nature would have been more accurate. Conservation had been ditched in favor of environmentalism.

In the 1980s, Nicholson used to warn volunteer activists, "I tell you never trust the state."[53] As the 1980s turned into the 1990s, WWF's links with governments deepened. Along with other NGOs, WWF became a political player working hand in glove with the state. By 1993, income from governments and aid agencies accounted for over half WWF-Germany's income.[54] At the same time, NGOs got into the business of monetizing their brands. When Kleenex developed its "Wildlife Tissues," sales leaped 76 percent.[55] "Green consumerism is a target for exploitation," Jonathon Porritt of Friends of the Earth complained. "There's a lot of green froth on top." Business, Fred Pearce wrote, "was putting the con into conservation."[56] It was only a beginning. The selling out of nature at the end of the 1980s and into the 1990s was trivial compared to what would follow.

* WWF-US stuck with wildlife. According to its head, Bill Reilly, American members loved wildlife. "Foreign aid, biodiversity, international development—we had polled on these and nobody marched to them in our sphere." (Alexis Schwarzenbach, *Saving the World's Wildlife: WWF—The First 50 years* [London, 2011], p. 179.)

17

Swallowing Hard

> I can't believe that science and technology leave us better off. The real wealth is nature, which science and technology are destroying.
>
> *Teddy Goldsmith, editor of* The Ecologist[1]

In 1999, Germany's largest environmental organization began exploratory talks with Europe's largest carmaker. NABU—the Nature and Biodiversity Conservation Union and its 560,000-strong membership—is Germany's equivalent of the Sierra Club. With 2014 revenues of $268.6 billion, Volkswagen AG is Germany's largest company, ranking number eight in the Fortune Global 500.[2] Agreement was reached the following year. One of its first fruits was a joint workshop on Volkswagen's power-train technology and fuel strategy. A year after it began, the federal government's Board on Sustainable Development acclaimed the NABU–VW partnership as "best practice" in its National Sustainability Strategy. Subsequent projects included organizing organic food lines at Volkswagen canteens, a Welcome Wolf! project to reintroduce wolfpacks into Saxony (Volkswagen's headquarters is in Wolfsburg, Lower Saxony), and participation in the NABU "Switching—saving fuel easily" campaign.[3] NABU burnished VW's eco-credentials. At the 2012 Geneva motor show, it proclaimed VW would be the most ecological car manufacturer in the world by 2018.[4] In return, the car maker's financial arm agreed to provide €1 million to a newly founded moorland protection fund, and in 2014 it paid €530,000.[5]

VW wasn't the only one. Daimler also paid NABU €920,000 to fund moorland projects that, perhaps not uncoincidentally, removed environmental objections to building a test track in a former military training area in the Black Forest. A payment of half a million euros to

NABU silenced objections to a new wind farm in the Hessian Vogelsberg, an important route for migratory birds. Opposition to construction of an offshore wind farm in the Wadden Sea, a UNESCO world heritage site off the coast of Denmark, Germany, and the Netherlands, dissolved after the developer paid €800,000 to BUND Friends of the Earth Germany.[6] Greenpeace also got in on the act. Having attacked "Volkswagen—Das Problem" in one of its key 2012 campaign messages, at the beginning of 2013, Greenpeace abruptly changed tack. Now they were best of friends. "Volkswagen and Greenpeace affirm strict CO_2 limits for new cars," limits that were only going to be met through further dieselization and worsening urban air pollution but showing the effectiveness of what is known in Sicily as a *pizzo*, a protection racket.[7]

"Isn't this sort of cooperation ultimately just Greenwashing?," asked a NABU–Volkswagen brochure in April 2014. "Those who practice greenwashing do so to distract from their dirty core business. In contrast to that, Volkswagen has committed itself to becoming the "leading ecological automotive manufacturer worldwide."[8] "Power-train technology," "fuel strategy," and "fuel saving" are clues to what was really going on—greenwashing the push by Europe's largest car manufacturer for the dieselization of Europe's automobile fleet.

The VW scandal broke on September 18, 2015, when the EPA issued Volkswagen a notice of violation alleging that its diesel cars had software designed to circumvent EPA emissions standards. "Using a defeat device in cars to evade clean air standards is illegal and a threat to public health," EPA assistant administrator Cynthia Giles said.[9] Bad as that was—and it was very bad—there was a still bigger scandal sitting behind the VW cheat software. By 2011, over half of all new cars sold in Europe were diesel.[10] The EU had deliberately engineered a massive switch from petrol- to diesel-driven cars in full knowledge that diesel exhaust is far more polluting and detrimental to urban air quality.

The EU's dieselization program sprang from the search for energy efficiency in the wake of the 1997 Kyoto Protocol. Compared to petrol engines, diesel held the prospect of improved fuel economy and lower carbon dioxide emissions per gallon of fuel, a proposition strongly supported by the IPCC.[11] There was a huge flaw in the miles per gallon analysis; depending on fuel quality, diesel contains 14 percent more carbon per gallon. A 2013 analysis of the EU dieselization policy concluded that any carbon dioxide emission advantages of diesel cars are trivial when measured against modern petrol-driven engines.[12] What Europeans got in exchange for negligible reductions in harmless carbon dioxide vehicle emissions was worse air pollution in their towns and cities, facilitated by the EU setting emissions standards that permitted diesels to legally emit ten times more nitric oxide and nitrogen dioxide (NOx) than petrol-driven cars.[13] To push car buyers to diesel, the EU encouraged member states to tax diesel less than petrol.

The fixation with carbon dioxide emissions meant downplaying the pollutants emitted from car tailpipes. According to a 2011 study, diesel engines emit 20 to 40 times more NOx than similarly sized petrol-engine cars, which explains the pungent, chlorine-like odor of diesel exhaust.[14] Of particular concern for public health is PM-10, microscopic particulates measuring ten micrometers or less. In June 2012, the World Health Organization upgraded its 1998 classification of diesel exhaust from probably carcinogenic (Group 2A) to Group 1, based on "sufficient evidence that exposure is associated with an increased risk of lung cancer."[15] As a result of the EU's diesel push, the improving air quality trend in European cities is being reversed. Paris, London, and Florence are among the cities with the highest levels of NOx. Since 1996, Luxembourg, Europe's diesel car capital, has experienced a 30 percent increase in NOx levels.[16]

The consequences of putting more diesel cars on the road were entirely foreseeable and foreseen. A 1993 British government report on diesel vehicles warned of their detrimental impact on urban air quality. "Any increase in the proportion of diesel vehicles on our urban streets is to be viewed with considerable concern."[17] Writing in the *Sunday Times* shortly after the VW scandal broke, political columnist Dominic Lawson quoted a senior Department for Transport civil servant involved in formulating the prodiesel policy.

> We did not sleepwalk into this. To be totally reductionist, you are talking about killing people today rather than saving lives tomorrow. Occasionally we had to say we were living in a different world and everyone had to swallow hard.[18]

It was a world created by environmentalism and carbon policy monomania.

By then, the extent of the EU's wind and solar renewables policy disaster had become apparent. In January 2014, the European Commission published its post-2020 climate and energy proposals. Progress toward the 2020 target had not been an unmitigated triumph, and EU firms and households were increasingly concerned by rising energy prices and widening energy cost differentials with the United States, the commission noted. It came out against biomass policies, questioning their ability to cut greenhouse gas emissions, and acknowledged that member states had overincentivized investment in wind and solar PV. More market-oriented approaches were needed "to the greatest extent possible."[19] Most decisive of all, the commission decided that its post-2020 27 percent renewables and 40 percent carbon dioxide reduction targets should not be binding on member states. In effect, the commission was running up the white flag on its renewable energy policy in a way that as few people as possible would notice. A former senior commission official, Steffen Skovmand, explained the legal position in a letter to the *Wall Street Journal:**

> Since the commission has decided not to propose binding targets, there will be no binding targets in the future because legislation within the EU originates from the commission alone. . . . If member states are no longer to be bound by concrete targets and goals, then no one is. The forty percent "target," in other words, is nothing but hot air.

He went on to note that Germany and Spain had suffered most from the EU's "disastrous utopian policies."

> The commission can not admit as much—that is not the way the music plays politically anywhere and certainly not in Brussels. But make no mistake, this commission communication signals a real EU turnaround.[20]

Skovmand erred in one respect. Germany was not a victim of EU policy; the EU had been a victim of German politics. In 2007, Angela Merkel had knowingly sabotaged Europe's cap-and-trade Emissions

* Skovmand's letter had been provoked by an article I had written in the *Wall Street Journal* saying that, institutionally, the EU had no reverse gear. "Like Frankenstein, the EU has created a renewable energy monster it does not know how to tame," I argued. ("Europe's Renewable Reversal," *Wall Street Journal*, January 29, 2014.)

Trading Scheme when she pushed EU leaders to adopt the 2020 renewables target. Her ability to do so reflected Germany's preeminent reputation in Europe. Since Hitler's defeat, Germany had accrued great international prestige thanks to the strength of its economy, its engineering prowess, and a succession of impressive political leaders—Konrad Adenauer, for rehabilitating West Germany in the Western family of nations and initiating the Franco-German friendship; Ludwig Erhard for the German economic miracle; Brandt for his Ostpolitik; Schmidt for being an island of stability in a turbulent decade; Kohl for his partnership with Reagan in ending the Cold War and reunifying Germany; and Schröder for his tough economic reforms.

The rise of the Greens brought about a profound transformation of German politics. Championing eco-ideologies buried in the rubble of 1945 and subsequently driven to the margins of German politics, the Greens had a creeping impact in replacing the pragmatism and realism of postwar Germany with an irrationality that is a marked feature of earlier and unhappier episodes in German history. It was a development that took mainstream German politicians unawares. As the EEG renewables law was going through parliament during the Red–Green coalition's first term, almost no one realized the Pandora's box of green subsidies that was being opened. Although the energy minister, Werner Müller, opposed the law, among Schröder's staff the opinion was, according to the journalist Alexander Wendt: Well, that's Scheer and his three other supporters—let the propeller heads have what they want, it can't be that expensive and will only help make the Greens more loyal partners.[21]

Merkel had less excuse. Whereas Schröder had trained as a lawyer, Merkel had studied physics and physical chemistry at East Germany's most prominent institute, the Akademie der Wissenschaften der DDR, and in the Soviet Union. She understood the destructive consequences of the rapid build-out of wind and solar PV but pressed ahead for reasons of political advantage—then inflicted it on the rest of Europe. Merkel embodies the second transformation Germany underwent—absorption of East Germany and its political culture rooted in systematic deception and manipulation, where the truth could not be spoken, people did not say what they really thought, and family and friends might well be Stasi informers. Born and raised in a culture of political repression, it would be more surprising if Merkel the politician had emerged from East Germany completely immune from it than if her political character had been formed by it.

Germany's *Energiewende* has done no one any good apart from lining the pockets of rent seekers. It retarded the quest for innovation to develop more efficient generating technologies. The EEG's promotion of expensive and inefficient technologies such as solar crowded out cheaper alternatives, according to a 2009 study. The EEG support mechanism, the study concludes,

> is a classic example of an unsound energy policy that is highly prone to lobbyism. It is very unlikely that such government-directed programs, picking winners and losers, would yield a more efficient energy mix than what would be determined in the market absent massive government intervention.[22]

Then there's what Wendt calls "Das CO_2 Paradox." As we saw in Chapter 13, more wind and solar PV capacity has not been matched by lower carbon dioxide emissions (they fell slightly in 2014 thanks to a mild winter)—a paradox only if it is assumed that the Greens' objective is to cut greenhouse gas emissions. If reducing CO_2 emissions had really been the priority, Germany would be building zero-emission nuclear power stations and championing hydraulic fracturing—fracking—of shale gas. Instead Germany is closing its nuclear power stations and has banned fracking. Low-carbon-dioxide-emitting CCGT plant is being closed while higher-CO_2 lignite-burning power stations have been built. Between 2011 and 2015, Germany brought online 10.7 GW of new coal-fired capacity. Their annual output will far exceed that of existing solar panels and will be approximately that of existing solar PV panels and wind turbines combined.[23] Closing nuclear power stations incurs a big CO_2 penalty. Not that it worries Germans. A 2013 poll by the Forsa Institute found that 47 percent of Green voters support the use of lignite.[24] Actual behavior reveals that stopping global warming is a pretext, not the reason, for Germany's pursuit of renewable energy.

There is a pattern that sheds light on the apparent inconsistency of the values driving those behaviors. Environmental concern was ignited when forests were purportedly threatened by acid rain from coal-fired power stations, but not when forests are ripped up to make way for wind farms. There is alarm about alleged public health impacts of NOx emitted from coal-fired power stations, but not about the health impacts of infrasound (very-low-frequency sound waves) caused by rotating wind turbines inducing motion sickness–like nausea in a minority of individuals, which have been known about for 30 years.[25] The postwar environmental movement came into being with the DDT scare ("A Fable for Tomorrow" is the first chapter in *Silent Spring*) but is silent over the IPCC-sanctioned eco-slaughter of raptors and migratory birds.

This pattern indicates the partial nature of environmental concern: It depends on the cause, not the effect. Green values elevate agricultural modes of energy production (harvesting wind and sun from wind and solar farms) over industrial forms, especially coal-fired power stations. For the environmental movement, there is one thing worse than coal. Hostility to nuclear power is a leitmotiv of environmentalism. Nuclear power symbolizes technology's reckless release of vast amounts of energy, leaving highly radioactive residues. For environmentalists, myth is stronger than reality. Nuclear power has relatively little impact on habitats, Hambler and Canney write in *Conservation*. Surprising as it might seem, the nuclear tests on Bikini Atoll helped preserve outstandingly good, undisturbed coral reef communities that are refuges from fishing.[26]

Nuclear power provides a test of climate change policy bona fides. Across the Rhine, in July 2015 the French National Assembly passed an Energy Transition for Green Growth Bill to cut the share of nuclear in France's power mix from 75 to 50 percent. The law made good a 2012 election pledge borne of political desperation by a nervous François Hollande aimed at winning support from the French Greens. Hollande then let the Greens have leading positions in 17 parliamentary constituencies, seats they would not otherwise have won. Even so, the Greens' 17 seats paled next to the left's 102-seat majority. Unless nuclear is used for balancing the influx of intermittent renewables, it will suck in more hydrocarbon-generated electricity. Signing the bill into law in August, Hollande quoted the surrealist poet Paul Eluard, "There is another world, but it is in this one," revealing the host of the December 2015 Paris climate conference to be if not a dreamer then the world's number two climate change hypocrite after Angela Merkel.[27]

During this time, America also experienced an energy transition. Unlike Europe's, it has been driven by technology (fracking) and market forces (plunging natural gas prices), not by government. Over the same period (1999–2012) that Germany *increased* power station emissions by 17.2 million metric tons, American power stations *cut* theirs by ten times that amount—170.1 million metric tons. The market-driven switch to natural gas saw annual carbon dioxide emissions from power stations burning coal and oil fall by 435.5 million metric tons, while CO_2 emitted from generating electricity from natural gas rose by 232.6 million metric tons. Indeed, the fall in America's annual emissions over those 13 years was very nearly half the *total* emitted CO_2 by all of Germany's power stations.[28] These reductions were not the product of governmental intent. They were an unanticipated byproduct of capitalism's ceaseless quest for profit by finding and making stuff cheaper.

Even before President Obama and the EPA decided that America should forsake its market-driven energy transformation, its largest state had embarked on its own *Energiewende*. Much of its inspiration was drawn from European ecological thinkers. With an admixture of homegrown anticapitalist ideologies, California became a laboratory of potent progressive cocktails. They would, in due course, be served up to the rest of the country when "No, thank you" would not be taken for an answer. But there was nothing they could do to alter the fundamental physics of electricity generation that had condemned Europe's attempt to failure. The only difference would be that the EPA and the American *Energiewenders* already had the example of Europe's failure before them, making theirs an even greater failure of policy.

18
Golden to Green

> The earth does not belong to mankind, according to Paul: "For the Earth is the Lord's, and the fullness thereof."
>
> *Arne Næss*[1]

> A Buddhist economy would make the distinction between "renewable" and "non-renewable resources." . . . The former co-operates with nature, while the latter robs nature. The former bears the sign of life, while the latter bears the sign of death
>
> *Fritz Schumacher*[2]

> The power of the Clerisy stems primarily not from money or the control of technology, but from persuading, instructing, and regulating the rest of society.
>
> *Joel Kotkin*[3]

German energy policies, French labor regulation, Italian public debts, and a Scandinavian cost of living premium: Subtract Silicon Valley and Hollywood, and America's most populous state has distinctly European features. California had been a pioneer in the preservationist movement dating back to John Muir and the founding of the Sierra Club in 1892, and for the better part of a century California had it both ways. Preservation of the Yosemite Valley, the Sierra Nevada, stretches of the coastline, and its redwoods did not put a brake on the state's booming economy.

An oil-well blowout six miles off the coast of Santa Barbara in 1969 began to change the balance between growth and environmental protection. Although the long-term environmental impacts of the spill were minimal, it led to the creation of the Californian Coastal Commission and an indefinite moratorium on offshore drilling within three miles of the coast. Topography also played a role in the development of Californian ecofundamentalism. Geography and weather patterns give Southern California its sunny climate and some of America's worst urban air quality. Cold ocean currents and nearby mountains lead to frequent temperature inversions, when warm air traps cooler—often polluted—air. Meeting the standards prescribed in the 1970 Clean Air Act presented California with insuperable problems: Pollution from trucks and autos means that the big cities of Southern California could be completely deindustrialized and still suffer from poor air quality.

Something else was in the air in 1975, when Jerry Brown succeeded Ronald Reagan as governor. Reagan had supported economic growth and conservation, backing tougher auto emission standards, blocking development around Lake Tahoe and extending state parks. Brown, in the words of Manhattan Institute senior fellow Steven Malanga, proselytized radical antigrowth environmentalism drawn from ecological thinkers such as Norwegian philosopher Arne Næss and the author of *Small Is Beautiful*, Fritz Schumacher.[4] Næss, a member of the resistance during the Nazi occupation of Norway, characterized conservation programs as "shallow ecology," proposing instead a deep-ecology rollback of industrial development. Like Næss, Schumacher had been an anti-Nazi, fleeing Germany for Britain, where his reputation (and the royalties from his book of Buddhist economics, as he referred to *Small Is Beautiful*) disinfected the organic Soil Association movement of its founders' pro-Nazi sympathies.*

Californian universities became seminaries of green ideologies raising up a secular clerisy, to borrow Joel Kotkin's term, to propagate their values through the media and into the governing bureaucracies of the state. From 1965, Herbert Marcuse was teaching the Frankfurt School's critical theory at the University of California, San Diego; Paul Ehrlich, author of the 1968 bestseller *The Population Bomb*, which predicted millions of deaths from mass starvation, had been

* The origin of the Soil Association within a pro-German, English landowning circle is told in Rupert Darwall, *The Age of Global Warming—A History* (London, 2013), pp. 39–41.

teaching biology at Stanford since 1959; and Garret Hardin was teaching his antihuman population theories at the University of California, Santa Barbara throughout the 1960s. In 1969, Friends of the Earth was founded in San Francisco by the former executive director of the Sierra Club and vocal advocate of population control, David Brower, with, as we've already seen, oil-industry funding.

All this helped bring about a profound transformation in Californian politics. According to Kotkin in his 2014 book *The New Class Conflict*, the old plutocracy—notably energy, manufacturing, mass agriculture, and construction—generally supported the economic advancement of the classes below them. The consensus across the political spectrum in America that growth was good remained universal at least until the late 1960s.

> For all its many environmental and social shortcomings, the old economic regime emphasized growth and upward mobility. In contrast the new economic order focuses more on the notion of "sustainability"—so reflective of the feudal worldview—over rapid economic expansion.[5]

In California, it is being replaced by a new, deeply stratified social order. As Kotkin puts it, the oligarchs of a dematerialized economy who made billions out of information technology (IT), finance, and entertainment have shaped a new kind of postindustrial economy. Its lighter environmental footprint becomes a license to deny those less well adapted or unfortunate a share in its riches and its lifestyle.

Antipollution regulation accelerated California's deindustrialization. Aerospace manufacturing shifted to states with lower energy costs and right-to-work laws. In the ten years to 2015, Southern California's industrial base shed 60 percent of its workforce, from 900,000 to 364,000.[6] Environmental progress is socially regressive in its application, Kotkin argues. California has America's largest number of billionaires (111) and its highest poverty rate (23 percent) on the Supplemental Poverty Measure. With roughly 12 percent of America's population, California accounts for roughly one-third of its welfare recipients.[7] The clerisy's no-growth planning policies increase downward mobility, particularly among Latinos. Shockingly, native-born Latinos have shorter life spans than their parents.[8] Without broad-based economic growth, large parts of an emergent middle class risk becoming a permanent class of low-wage proletarians, Kotkin suggests.[9] Technology replaces religious faith or civic virtue to provide a secular justification for increased social stratification in what Kotkin dubs "high-tech feudalism."[10] The proletarianization of the American middle class is not an unfortunate or unwelcome by-product of green ideas. Killing the American Dream is necessary for the success of the Europeanization project because ever-rising material consumption—especially the burning of hydrocarbons—is the clerisy's biggest prohibition.

California's social stratification has a geographic dimension. The elites that run the state live within five miles of the coast, where air conditioning is a renewable resource provided by the cool waters of the California current. Rather like apartheid-era South African townships, the impoverished lower-middle class live out of sight and out of mind a hundred miles away, in the sweltering interior. When it comes to electricity bills, it pays to be wealthy. According to Kotkin, the average summer electrical bill in rich, liberal Marin County was $250 a month, while in poorer, hotter Madera in the San Joaquin Valley, the average bill was twice as high.[11] Google chairman Eric Schmidt says people in Silicon Valley don't talk about the concerns of the 99 percent because a lot of them are immune to those concerns. "We live in a bubble, and I don't mean a tech bubble or a valuation bubble. I mean a bubble as in our own little world," Schmidt acknowledged.[12]

As we will see, Schmidt's and the valley's indifference to the majority of Californians was manifested in their efforts to raise electricity costs by imposing wind and solar energy mandates and then defeating the 2010 Proposition 23 ballot initiative to partially reverse them. California's path to energy ruin has been three decades in the making. Thanks in large part to green antigrowth ideology, hardly any new generating capacity had been built in the last two decades of the twentieth century. Between 1990 and 1999, California's generation capacity decreased by 2 percent while consumption increased by 11 percent. Aging generators, a diminishing capacity margin, and aggressive environmental regulation set the stage for California's 2000–2001 energy crisis.

Toward the end of 2000, a significant amount of generating capacity was undergoing maintenance. Low water levels in the Pacific Northwest reduced hydroelectric output, and several gas-fired plants had used up their pollution credits and would be liable to fines if used. Preserving grid stability therefore necessitated rotating blackouts. The capacity squeeze also caused a huge price spike. By December 2000, wholesale prices were 11 times higher than a year earlier. To make matters worse, the state imposed retail price caps so electrical utilities couldn't recover higher wholesale costs. Between June 2000 and April 2001, when it filed for Chapter 11 bankruptcy protection, the Pacific Gas and Electric Company incurred $9 billion in costs that the state prevented it from recovering.[13]

Even so, California carried on down the path of energy incoherence. Between 2002 and 2014, coal, natural gas, and nuclear power station capacity—the most reliable generating technologies—fell by 8.5 percent while hydro capacity nearly halved. Overall, the proportion of non–weather dependent generating capacity fell from 95 percent of system capacity in 2002 to 80 percent in 2014. Over the same period, total generating capacity shrank nearly 9 percent but demand rose 27 percent. To keep the air-conditioning on, California more than doubled the electricity it imported from neighboring states in 2014, accounting for one-third of Californian demand.[14]

Politicians in Sacramento competed with each other to save the planet from global warming—and enrich wind and solar magnates. In 2002, the state legislature passed the California Renewable Portfolio Standard Program, setting a 20 percent renewables target by 2017. In 2003, it was brought forward to 2010 and put into law in 2006. In 2005, Governor Arnold Schwarzenegger proposed raising the target to 33 percent by 2020 and issued an executive order setting three greenhouse gas reduction targets: Back to 2000 levels by 2010—a target that was met thanks only to the 2008 recession; to 1990 levels by 2020; and 80 percent below 1990 by 2050. Silent on the means of meeting the targets, Schwarzenegger wasn't silent on who would benefit: "California companies investing in these technologies are well-positioned to profit from this demand."[15] The targets became law with Assembly Bill 32, the California Global Warming Solutions Act of 2006.

As the damage caused by AB32 became better understood, spring 2010 saw the launch of a campaign to get the nearly half a million signatures necessary for a ballot initiative to suspend AB32. The initiative succeeded and became Prop. 23 but was defeated in that November's general election by 61.5 percent to 38.5 percent. Flushed with success, in 2011 the legislature wrote Schwarzenegger's 33 percent target into law, by which time Jerry Brown was back in the governor's mansion for a third term. The definition perversely excluded nuclear power and large hydro facilities. Like in Germany, tackling global warming is not the objective but an excuse to convert the grid to wind and solar. In January 2015, Brown upped the ante with the goal of a 50 percent renewables target by 2030. It was, the newly inaugurated fourth-term governor declared,

> exactly the sort of challenge at which California excels. This is exciting, it is bold and it is absolutely necessary if we are to have any chance of stopping potentially catastrophic changes to our climate system.[16]

Zero-emissions nuclear power wasn't mentioned. Too exciting, too bold, too logical.

The crushing of Prop. 23 was a demonstration of the raw political power of environmentalism and California's elites. Schwarzenegger had slammed arguments made by Texan oil refiners that suspending AB32 would create jobs as like "Eva Braun writing a kosher cookbook."[17] The Californian establishment was mobilized. George Shultz was co-chairman of pro-AB32 Californians for Clean Energy and Jobs group. Conceding that some companies were worried about the cost of AB32, "the new regulations will boost the state's economy by creating 'clean-tech jobs,'" claimed Ronald Reagan's former secretary of state and Richard Nixon's director of Office of Management and Budget.[18]

What AB32 supporters lacked in intellectual edge, they more than made up with an overwhelming money advantage. They cried foul about out-of-state oil money when the real story was the money pouring in behind AB32 to defeat Prop. 32. The Rockefeller Family Fund chipped in $300,000. In a display of gratitude to the legislators who had bankrupted it, Pacific Gas and Electric Company contributed half a million dollars. The usual suspects were unusually generous: the National Wildlife Federation ($3m); the National Resources Defense Council ($1.9m); the Sierra Club ($1.7m); the League of Conservation Voters ($1.3m); the Environmental Defense Fund ($1.1m); Nature Conservancy ($800,000); the Union of Concerned Scientists ($113,005); and the National Audubon Society ($100,000).[19] The big bucks in the defeat of Prop. 23 came from eBay and TechNet, a tech lobbying group including Apple, Google (Wendy Schmidt, Eric's wife, gave $500,000), Yahoo, Silicon Valley venture capitalist John Doerr ($2.1m), and billionaire Vinod Kholsa, formerly of Sun Microsystems ($1.0m).[20] The biggest anti-Prop. 23 donor was hedge-fund manager Tom Steyer ($5m), who co-chaired Californians for Clean Energy and Jobs along with Shultz.

These donor numbers disguise the upstream source of much of the anti-Prop. 23 NGO funding. A 2013 report by the Columbia School of Journalism notes a sharp rise in environmental funding from a handful of primarily West Coast funders. This followed publication in 2007 of a strategy report, "Design to Win," commissioned by six wealthy foundations.[21] Written by researchers from California Environmental Associates and the Stockholm Environment Institute, the report argued that solving global warming required "a makeover of the global economy that is unprecedented in both scope and speed." Philanthropy could play a pivotal role in bringing this about, but "donors and foundations must be strategic and choose interventions with the most potential to set the world on a low-carbon course."[22]

The report's message was heeded. In 2008 three of the wealthiest foundations in America—the William and Flora Hewlett, David and Lucile Packard, and McKnight foundations—committed more than $1.1 billion to launch ClimateWorks. Hewlett alone pledged $500 million, making it the single largest grant in the foundation's history.[23] ClimateWorks—"our mission is to mobilize philanthropy to solve the climate crisis," according to its Twitter bio—directs the flow of foundation and hedge-fund dollars to maximize their impact.[24] Foundation dollars donated to ClimateWorks is rebundled and transferred to the secretive Energy Foundation, which in turn serves as a major source of grants to American NGOs.*[25] The mixing and aggregation of climate dollars also has the effect of hiding the upstream source of the cash originating, in the words of the Columbia Journalism School, from a "small cadre of wealthy hedge fund owners and foundations headquartered primarily in California."[26]

The strategy in the "Design to Win" report went beyond California and the United States. "The global community must overcome

* The Columbia Journalism School authors state that representatives of the Energy Foundation would not speak to them (p. 32).

the collective action problems that have hobbled international climate agreements," the report's authors argued. "A cap on carbon output—and an accompanying market for emissions permits—will prompt a sea change that washes over the entire global economy."[27] As well as putting $900,000 into its own backyard in the fight against Prop. 23, ClimateWorks funded the European Climate Foundation to lobby for steeper cuts in greenhouse gas emissions.

What the media report—and what the public sees—is the downstream NGO activity the foundations buy. A 2014 minority staff report of the Senate Committee on Environment and Public Works (EPW) identified ten large foundations active in promoting environmentalism with reported assets totalling $23.2 billion.[28] Between 2010 and 2013, the eight NGO donors to the anti-Prop. 23 campaign (excluding ClimateWorks) received a total of $202.2 million in grants, the top two recipients being Nature Conservancy ($58.6m) and the Environmental Defense Fund ($53.9m). John Podesta's Center for American Progress also received $8.4 million in grants from them.[29] NGOs don't have to be viewed as sock puppets for there to be something deeply troubling about the relationship between the visible activities of NGOs and their hidden dependence on a small group of hyperwealthy individuals and foundations. As the EPW minority staff observes, environmental NGOs that are heavily reliant on foundation funding for a substantial proportion of their budgets

> begin to look much more like private contractors buying and selling a service rather than benevolent nonprofits seeking to carry out charitable acts.[30]

Environmental groups portray themselves as guardians of undefiled nature and protectors of wildlife. Being in the pay of West Coast oligarchs means environmental groups can't be genuine advocates for wildlife. In 2006, the then president of the National Audubon Society, John Flicker, wrote an article on wind power that was to become notorious. Wind was a good-news, bad-news story. "The good news is that many new wind-power projects are being proposed across the country," Flicker wrote. "The bad news is that wind turbines sometimes kill a lot of birds." They certainly do. A 2013 study estimated 888,000 bat and 573,000 bird fatalities a year, including 83,000 raptors, not counting those killed by the extra power lines and pylons required to connect wind farms to the grid.[31] "On balance," Flicker continued, "Audubon strongly supports wind power as a clean alternative energy source that reduces the threat of global warming" before blaming Congress for incentivizing wind-power investors to cut corners by not making the wind-energy-production tax credit permanent.[32] "We very much appreciate Audubon's leadership on this issue," the executive director of American Wind Energy Association purred.[33] Was it money talking? Audubon would receive $11.2 million between 2010 and 2013 from the ten funders identified in the EPW minority report.[34] It would also contribute $100,000 to defeat Prop. 23 and keep covering California with wind farms.[35]

Solar, too, is a bird killer. The 377 MW Ivanpah solar project, in which Google invested $178 million and the U.S. government chipped in $1.6 billion of loan guarantees, uses 300,000 mirrors in the Mojave Desert to reflect solar rays onto three boiler towers. Birds flying across the 3,500-acre site risk having their feathers burned in the 800 degree Fahrenheit solar flux. Workers call them "streamers," reporting an average of one streamer every two minutes. A 2014 analysis by the National Fish and Wildlife Forensics Laboratory found that severe singeing caused catastrophic loss of flying ability, leading to death. Less severe singeing has led to impairment of flight ability and increased vulnerability to predators and reduced ability to forage. Ivanpah is an equal opportunity killer. Seventy-one different species were identified, ranging in size from hummingbirds to pelicans. "Ivanpah may act as a 'mega-trap' attracting insects which in turn attract insect-eating birds," the researchers thought.[36] (At the photovoltaic site they examined, bird fatalities resulted from impact trauma from birds flying into PV panels.) Commenting on the Ivanpah deaths, Audubon's renewable energy director for California said they were alarming. "It's hard to say whether that's location or the technology," Garry George told the Associated Press. "There needs to be some caution."[37] Caution? The precautionary principle isn't invoked by environmentalists when it comes to their favored forms of energy production and the mass slaughter of birds in eco-friendly ways.

By contrast, the American Bird Conservancy took the Department of the Interior to court to overturn its decision to give wind-energy companies 30-year permits to kill protected bald and golden eagles. The court found that federal authorities, who were joined in the suit by the American Wind Energy Association, had not followed federal rules when deciding to extend wind farms' permits-to-kill from 5 to 30 years. Research funded by the conservancy found that 30,000 wind turbines had been installed in areas critical to the survival of federally protected birds and that more than 50,000 more were planned in similar areas.[38] The conservancy is cautious about accepting wind and solar: "We strongly believe that renewable energy sources should not be embraced without question."[39] Throwing caution to the wind, Audubon, on the other hand, "strongly supports" wind.[40]

19

Capitalism's Fort Sumter

Great nations are never impoverished by private, though they sometimes are by public prodigality and misconduct.

Adam Smith[1]

In lobbying for wind and solar energy, environmental NGOs were faithfully reflecting the views of California's green oligarchs. As Mark Mills puts it, Silicon Valley believes that Moore's Law—that computer processing power doubles every couple of years—applies to renewable energy technologies and will make them economic.[2] A myriad of venture capital speeches and pitches analogized clean tech as an emerging revolution akin to the digital disruption in computing and telephony, Mills says. In over a decade, more than $25 billion was invested in clean-tech ventures, plus another $50 billion in federal grants and gifts. Nearly everyone lost money. Clean tech became a dirty word in most venture capital circles. The limitations of physics and the chemistry of energy impose fundamental constraints that are not analogous to the miniaturization that has driven the exponential growth of IT speeds. Making clean tech viable depends not on technology, which is amenable to results-oriented managements of Silicon Valley, but on fundamental scientific breakthroughs. As Mills points out, often these occur serendipitously and frequently from unexpected people and venues. They can't just be ordered up.[3]

There is another Moore's Law. The second one says the richer you are, the more likely you are to support green causes. In 2000, the eponym of the first law and cofounder of Intel, Gordon Moore, and his wife set up the Palo Alto–based Moore Foundation with five billion dollars and later would spend one million dollars on the anti–Prop. 23

campaign.[4] Climate change is ethics for the wealthy: It legitimizes great accumulations of wealth. Pledging to combat it immunizes climate-friendly corporate leaders and billionaires from being targeted as members of the top one-tenth of the top one percent. This signifies a profound shift in the nature and morality of capitalism. In the famous passage on the invisible hand in *The Wealth of Nations*, Adam Smith wrote of individuals who, intending their own gain, promote an end that was no part of their intention. "By pursuing his own interest he frequently promotes that of society more effectually than when he really intends to promote it."[5] Less well known are the two sentences that immediately follow:

> I have never known much good done by those who affected to trade for the public good. It is an affectation, indeed, not very common among merchants, and very few words need be employed in dissuading them from it.[6]

One can be pretty sure that if supporting renewable energy harmed their interests, they would, as Adam Smith suggests, drop it in a nanosecond. The acquisition of green virtue does harm everyone else, especially the least well-off, and represents an existential threat to the energy-intensive economies of the Midwest from the Great Lakes down to Texas. As Kotkin puts it, this as a conflict between an economy that makes tangible things and one that deals in the intangible world of media, software, and entertainment—you can add finance—thereby squandering America's unique advantage in being a powerhouse in both the material and nonmaterial worlds.[7]

This economic civil war fought by the two coasts against the American heartland is also a civil war within American capitalism, waged as part of environmentalism's war on hydrocarbon energy. Silicon Valley and the IT industry generally refute Schumpeter's prediction of the atrophy of the entrepreneurial function by salaried managers. It does accord with Schumpeter's larger prediction regarding the ultimate demise of capitalism and bears out his aphorism of capitalism paying the people that strive to bring it down.

Capitalist wealth had been used to fund the Frankfurt School decades earlier (Chapter 5). Disbursing the billions of dollars being raised to fund environmentalists in America's economic civil war has become a big business in its own right. The 2014 EPW minority staff report revealed how green grant making had become increasingly prescriptive and centrally coordinated. In 2011, around $1.13 billion of all foundation giving to environmental causes was made by members of the New York–based Environmental Grantmakers Association (EGA). Like the Energy Foundation, the EGA is a secretive organization—it refused to disclose the identities of its near 200-strong membership to Congress or the public—that coordinates grant giving to maximize the strategic impact of green dollars. West Coast foundations were three of the top four EGA donor members in 2011: the Gordon and Betty Moore Foundation ($134.4m); the David and Lucile Packard Foundation ($121m), and the William and Flora Hewlett Foundation ($53.4m).[8] One recipient is Earthjustice, a public-interest law firm ("Because the Earth needs a good lawyer") spun out of the Sierra Club, which litigates against fossil fuel companies and advocates on behalf of renewables investors, in a vivid example of the weaponization of Silicon Valley dollars.

Spontaneous antibusiness campaigns, rather like Soviet-era manifestations, turn out to be centrally funded too. Since 2006, activist Bill McKibben has led three campaigns: Step It Up, against coal-fired plants; 1Sky, for renewable energy; and 350.org, to "fight iconic battles against fossil fuel infrastructure" (e.g., the Keystone XL pipeline). In a 2010 article, McKibben boasted that

> with almost no money, our scruffy little outfit, 350.org, managed to organize what *Foreign Policy* called the "largest ever co-ordinated global rally of any kind" on any issue.

It turns out that McKibben's scruffy little outfits were front organizations. Far from being a bottom-up, grassroots campaign, an investigation by Canadian researcher Vivian Krause found that four grants accounted for two-thirds of 350.org's budget and that McKibben's three campaigns had received more than 100 grants totaling $10 million from 50 charitable foundations, over half coming from three—the two big Rockefeller funds and the Schumann Center for Media and Democracy.[9] Other donors to 350.org included ClimateWorks and the Tides Foundation.[10]

Tides is an especially significant organism in the ecosystem of anticapitalism. Since it was founded in 1976, Tides has funneled money from sub-billionaire West Coast liberals to fund progressive causes. It funded the nuclear winter conference satellite hookup with Moscow in 1983 (Chapter 10). According to Jarol Manheim, its grant giving emphasized the creation of an infrastructure to support progressive activism,

> whether in the form of anti-corporate or pro-social responsibility research, message construction, social networking, policy development, strategy formation, or recruitment.[11]

Progressive philanthropist and networker par excellence Joshua Mailman, whose Threshold Foundation merged with Tides, outlined the strategy in a 2002 essay:

> As social activists interested in the future of social change, it is often necessary to seek the straw that breaks the camel's back, the most effective way to leverage our small resources to make major change.[12]

The progressives' goal was to create a narrative of the People vs. Polluting Corporations. Murray Edelman explained the progressives' logic in 1988: "To define the people one hurts as evil is to define oneself as virtuous."[13]

In the philanthropic funding stakes, the left enjoys a structural advantage thanks to the tendency of conservative money to spawn progressive causes. The $19.7 billion of assets (2011) of the three West Coast environmental foundations adds to the firepower of longer established progressive foundations, many of which had been endowed by supporters of free markets. The Pew Memorial Trust had been endowed by Joseph N. Pew Jr. in 1948. By 2014, the Pew Charitable Trusts had $6.2 billion of total assets, but Pew himself had been an opponent of the New Deal and a bulwark of conservative positions within the Republican Party.[14] In its early years, his foundation supported conservative think tanks such as the Hoover Institution and the American Enterprise Institute. During the 1980s, Pew's political orientation was flipped so that, as Manheim explains,

> a foundation that was established by prominent advocates of free trade and limited government is now a leading funder of policy

initiatives that generally are opposed by advocates of those same positions today.[15]

Environmental organizations were prime beneficiaries of Pew's political reorientation. In 1998, the Pew Charitable Trusts established the Pew Center on Global Climate Change as a global warming think tank, which in 2011 would morph into the Center for Climate and Energy Solutions (known as C2ES). In 2001 a Pew trust made grants to the Sierra Club ($280,000), Friends of the Earth ($300,000), Earthjustice ($571,000), and the Environmental Defense Fund ($1.07m).[16]

Something similar happened to the John D. and Catherine T. MacArthur Foundation. MacArthur was a small-government businessman with an animus against environmentalists, who had opposed some of his real estate developments in Florida. His foundation's original 1970 deed states one of its purposes as supporting "ways to discover and promulgate avoidance of waste in government expenditures."[17] After MacArthur's death in 1978, his son waged a legal battle against the foundation, ousted most of the board, and turned MacArthur's mission into ways of supporting progressive causes. At the end of 2013, it had $6.3 billion of assets and in 2011 was one of EGA top ten donors to environmental causes, making $24.2 million of grants to environmental groups.[18] In 1982 it established the World Resources Institute (WRI) with a $15 million grant. One of WRI's first reports was on acid rain in the western states of the U.S. "The things that give the West its beauty are what makes it so vulnerable," one of the report's authors told *Science*.[19] The report recommended tighter emissions controls and limiting the use of cars.

Three decades later, WRI had become Washington's go-to global warming think tank, reflected in the size and sources of its income. Of its $51.6 million income in 2013, just $2.6 million (five percent) came from foundations (those included the European Climate Foundation, ClimateWorks, and the Energy Foundation), compared with $5.3 million from the U.S. government. In turn, the U.S. government's contributions were dwarfed by payments from European governments, which totaled $28.7 million—over half WRI's income that year—including €1.5 million ($1.9m) from the European Commission for "designing the 2015 global climate change agreement."[20] What were the Europeans buying—influence or expertise? Both, probably.

Environmentalism was the key that unlocked the fortress gates of capitalism. There is a lot of make-believe in Silicon Valley's espousal of renewable energy, which crosses the line into outright dishonesty when claims are made that its data centers rely 100 percent on intermittent wind and solar energy. There's also more than a dash of hypocrisy. Google's fleet of private jets based at San Jose airport burned the equivalent of 59 million gallons of oil between 2007 and 2013. "Of course, the wealthy of the past, and more traditional plutocrats today, also consume at a high level," Kotkin comments on Google's Gulfstreams and Boeings, "but they at least do so without lecturing everyone else to cut their consumption."[21]

Green billionaire philanthropy upends traditional notions of charitable giving to help those most in need. For sure, there is a certain amount of going through the philanthropic motions of looking after the neediest. Eric and Wendy Schmidt's foundation supports Oakland's People's Grocery, which calls itself "a leader in the evolving food justice movement" by growing food in inner cities. Tom Steyer and his wife, Kat Taylor, have pumped money into the San Francisco-based TomKat Charitable Trust, which funds organizations that envision "a world of climate stability, a healthy and just food system, and broad prosperity."[22]

Ethanol policy and the Renewable Fuel Standard (RFS) provided a test for the West Coast's billionaire philanthropists. If there's one thing worse for the poor than higher energy costs, it's spiraling food prices. As Kathleen Hartnett-White and Stephen Moore write in *Fueling Freedom*, bioethanol is a prime example of counterproductive and "ethically offensive" energy policy.[23] In 2007, Congress strengthened the ethanol mandate with the Energy Independence and Security Act. The following year, the price of corn rose from $2.50 a bushel to nearly $8 a bushel, driving up food prices because of the increased cost of feed grains for livestock and poultry. Ethanol now accounts for 40 percent of the U.S. corn crop. In a June 2015 address to the Food and Agriculture Organization in Rome, Pope Francis questioned the nonfood use of agricultural products for biofuels.[24] Even Friends of the Earth opposes it. The big West Coast foundations could have mobilized to fight King Corn. Instead they sat on their philanthropic billions, except for Tom Steyer, who supported ethanol in the war on the hydrocarbon economy.*

Foundation dollars have also been deployed to manufacture eco-consciousness among blue collars and minorities. The results have all the authenticity of folk dancers dressed in colorful ethnic costumes greeting a communist party dignitary. The phoniness of the Blue-Green Alliance was exposed in January 2012 when it came out against Keystone XL and the Laborers' International Union of North America (LIUNA) quit in disgust. When AFL-CIO president Richard Trumka said that the labor movement was divided over the project, LIUNA general president Terry O'Sullivan thundered:

> That is an understatement. That divide is as deep and wide as the Grand Canyon. We're repulsed by some of our supposed brothers and sisters lining up with job killers like the Sierra Club and the Natural Resources Defense Council to destroy the lives of working men and women.[25]

In September 2015, 21 Democratic assembly members of the Californian legislature—including 11 blacks and Latinos—crossed party lines to vote with Republicans to stop a bill requiring steeper cuts in greenhouse gas emissions. "Who does it impact the most? The middle class and low income folks," one of them shot back.[26]

Environmentalism fueled by West Coast billionaires and philanthropic foundations meant that working people lost the political party that was meant to represent them. Money can't buy me love, but it had bought the soul of the Democratic Party.

* In August 2014, Steyer's NextGen Climate praised the RFS: "The RFS supports 73,000 good-paying, clean energy jobs in Iowa and is helping us reduce our dependence on fossil fuels." Ben Geman, *National Journal*, August 19, 2014, http://www.nationaljournal.com/energy/2014/08/19/Tom-Steyer-Takes-Side-Environmentalists-Ethanol-Fight.

20

The Washington, D.C., *Energiewende*

> America is on the verge of technological breakthroughs that will enable us to live our lives less dependent on oil.
>
> *George W. Bush, 2007*[1]

> We were the leaders in wind. We were the leaders in solar. We owned the clean energy economy in the '80s. Guess what. Today, China has the most wind capacity. Germany has the most solar capacity. . . . We've fallen behind on what is going to be the key to our future.
>
> *Barack Obama, 2011*[2]

Elections have consequences. The 2006 midterms gave Democrats majorities in both houses of Congress. The next two years saw the three branches of government clear the ground on which President Obama's *Energiewende* would be built. Unlike Europe, America had already embarked on its own energy transformation, one not foreordained by politicians and regulators but brought about by the genius of capitalism and wrought by Adam Smith's invisible hand.

Geologists had long known about enormous shale formations such as the Bakken Shale of North Dakota and Montana, the Barnett Shale of Texas, and the Marcellus Shale running through northern Appalachia. Hydraulic fracturing dates back to the 1940s but was not widely deployed until 2003. Other than exempting fracking from the Safe Drinking Water Act in 2005, as far as politicians in Washington were concerned, the fracking revolution might as well have been happening on another planet. Instead they passed mandates and regulations and handed out billions in subsidies, accompanied by tired clichés of

Sputnik moments and Apollo missions aimed at ridding the world of hydrocarbon energy. All the while, fracking was reducing America's carbon dioxide emissions and doing what the politicians said they wanted without costing taxpayers and consumers a dime.

Washington, D.C.'s *Energiewende* kicked off with George W. Bush's 2007 State of the Union address. Standing in front of the first Democratic speaker in 12 years, Bush described a "great goal" of reducing gasoline usage by 20 percent in 10 years, which would unleash an anti-fracking counterrevolution: "We must increase the supply of alternative fuels by setting a mandatory fuels standard to require thirty-five billion gallons of renewable and alternative fuels."[3] It could have been Jimmy Carter—or Barack Obama. Alternative fuels and aggressive fuel economy standards were embodied in the bipartisan Energy Independence and Security Act, signed into law in December 2007. Bush's final State of the Union address, delivered a month later, enthused about trusting the creative genius of American researchers to pioneer a new generation of clean technology. "Our security, our prosperity, and our environment all require reducing our dependence on oil," Bush declared.[4] In saying that America was on the verge of technological breakthroughs to reduce its dependence on oil, the President left out a crucial word—"foreign"—which would otherwise have described the revolution that was already under way, with no thanks to him and the members of Congress he was addressing.

The transition from Bush to Obama was well-nigh seamless. "We have known for decades that our survival depends on finding new sources of energy, yet we import more oil today than ever before," Obama said in his first address to Congress, in February 2009. The new President promised to spend $15 billion a year on wind and solar power, advanced biofuels, clean coal, and technologies toward making road vehicles more efficient. Americans could breathe easier; so-called clean diesel, which Bush had pushed, didn't make the grade. Nuclear, another item on Bush's menu, hadn't either, but made a comeback in Obama's 2010 State of the Union address, with talk of building a new generation of nuclear power stations. By then, natural gas output had risen by 27 percent since the 2005 trough in natural gas output, which had seen the lowest volume of natural gas since 1992.[5] America's spontaneous energy transformation was reducing energy imports and greenhouse gases, which federal policies weren't. Corn-based ethanol supplied less than 5 percent of transportation energy and reduced U.S. oil imports by, at most, just 1 percent.[6]

Natural gas rated a grudging mention in Obama's 2011 State of the Union, as in "we need them all." All of the above didn't include oil. "I'm asking Congress to eliminate the billions in taxpayer dollars we currently give to oil companies," he said as a prelude to his big energy policy announcement: By 2035, 80 percent of electricity should come from "clean energy sources."[7] It would have been politically perverse in an election year not to have given America's capitalist energy transformation more than a passing mention. With natural gas output already up 14 percent and domestic oil production up 13 percent during the Obama presidency, Obama's 2012 address had the sound of the penny dropping.[8] Nearly ten years after first being widely deployed and five years after Bush had first started the ball rolling on Washington, D.C.'s *Energiewende*, fracking had forced itself to politicians' notice. "Right now—right now—American oil production is the highest that it's been in eight years. That's right, eight years," Obama boasted. "Not only that, last year, we relied less on foreign oil than in any of the past sixteen years."[9]

Opening up vast domestic reserves of hydrocarbons killed off the national security justification for energy policy that went back to Jimmy Carter's "moral equivalent of war" and Nixon's Project Independence speech nearly 39 years earlier and all the way back to 1945 and FDR's oil-for-security deal with the House of Saud.* So Obama had to come up with a new variant to justify policies that fracking was making redundant: America had only two percent of the world's oil reserves and there still wouldn't be enough oil to go around. As a rationale, it hasn't worn well. Just four years later, with the oil price at a 12-year low, the International Energy Agency was warning that the oil market risked drowning in oversupply.[10]

Nonetheless, it was high time for Obama to claim the credit. Over the course of 30 years, public research dollars had helped develop the technologies to extract natural gas from shale, "reminding us," Obama asserted, "that Government support is critical in helping businesses get new energy ideas off the ground," before contradict-

* For Nixon and Carter's energy policies, see Rupert Darwall, *The Age of Global Warming—A History* (London, 2013), pp. 82–86.

ing himself by complaining that subsidizing oil companies had gone on too long.[11] Federal funding of fracking research had been so important that Obama's 2012 address was the first to mention the word "shale" since Gerald Ford's "the state of the union is not good" address in 1975, which set a target of producing one million barrels of synthetic fuels and shale oil production per day by 1985.* As well he might. Fracking was the one bright spot in Obama's otherwise anemic economic record. According to a 2014 report by Mark Mills, fracking propelled the United States into being the largest and fastest-growing producer of hydrocarbons in the world. In just two years, U.S. oil production grew more than it had declined over the previous twenty. Natural gas was so abundant that the United States had gained a permanent competitive advantage for domestic industries and exports, Mills wrote.[12] Nearly 300,000 direct oil and gas jobs had been created since 2003. The $300 billion to $400 billion of annual economic gain from the oil and gas boom had been greater than the average annual GDP expansion of $200 billion to 300 billion.[13] Without the unplanned expansion of oil and gas, the economy would have continued in recession and Obama's 2012 reelection chances would have looked very different.

While politics led Obama to embrace fracking, environmental activists not facing tough election fights reacted to it as if it were the spawn of the devil. That the world was running out of oil—despite evidence to the contrary—was a deeply held article of faith needed to justify renewable energy. Their guru, Fritz Schumacher, had hailed the 1973 oil price shock as heralding the end of an era. World oil production then rose by 12 percent in 6 years. From the end of the 1990s, oil began to reverse a near two-decade trend of declining prices. In 2008, oil prices reached an inflation-adjusted peak not seen for 28 years, triggering a renewed bout of speculation on the imminence of peak oil. Now fracking had come along to make abundant what should have been scarce. Environmentalists mobilized. Eric and Wendy Schmidt's foundation teamed up with Tides and the Ithaca, New York–based Park Foundation to pour money into grassroots antifracking campaigns to get statewide fracking bans in New York and Colorado. In this way,

* In any case, Ford was referring to oil shale, or kerogen, which is different from the shale oil being produced from fracking in the Obama years.

shareholder wealth generated from Google colluded in an attempt to destroy more jobs and GDP than Google itself had ever created.

Park's president is the Duncan Hines and Park Communications heiress Adelaide Park Gomer, and the Park Foundation funds sustainable community initiatives in Upstate New York. Park money, often alongside funding from the Schmidt Family Foundation, was funneled into the New York–based Sustainable Markets Foundation, which then bundled cash out to activist organizations such as Water Defense, Frack Action, and Yoko Ono's Artists Against Fracking.[14] Park was also the key funder of Josh Fox's antifracking *Gasland* movies.[15] Receiving a 2011 Common Cause advocacy award, Gomer, who is also a board director of the Greenpeace Fund, described fracking as "one of the greatest environmental atrocities of our generation." Flying home from Philadelphia one evening, she was horrified to see that

> a once beautiful wilderness area was now dotted and lit up by blinding white lights and eerily disturbing gas flares. . . . I felt somewhat like an Iraqi must have felt in reaction to "shock and awe."[16]

Gomer complained that New York State's 1,500-page draft impact assessment on fracking had mentioned cancer only ten times. "I am concerned about the shockingly rising rates of environmentally caused cancer in our population," Gomer said, repeating the baseless claim from *Silent Spring* that, as we saw in Chapter 4, had in turn been derived from Nazi medicine. Fracking involved pumping hundreds of chemicals into the bowels of the earth. It would ruin the state's pristine landscapes, tourism, and the wine industry and contaminate organic vegetables being grown in Tompkins County, the Marie Antoinette of Upstate New York declared. "Nothing short of a total ban can save us from this unfolding tragedy!" Gomer told the Common Cause banquet.

> We believe that New York must become the first state to ban fracking taking a leadership role that the rest of the country can then rally behind.[17]

After months of procrastinating, in December 2014, Governor Andrew Cuomo folded and announced a total ban on fracking in the state. A N.Y. Department of Health study hadn't found any threats to public health but said there were risks that remained unanswered, perfectly illustrating the regressive nature of the precautionary principle. Instead, Cuomo backed plans for three new Las Vegas–style casinos. The Sierra Club was exultant. "This move puts significant pressure on other governors to take similar measures to protect people who live in their states."[18] None did, not even Jerry Brown.

Although losing the fight to stop fracking everywhere else, a new front had unexpectedly opened up thanks to NGOs and their billionaire backers. Democrats lost control of the House of Representatives in the 2010 midterm elections and held a precarious majority in the Senate. Given that regulation of electricity utilities was a matter for states, how could the federal government require that 80 percent of America's electricity come from renewable energy, as Obama had pledged in his 2011 State of the Union address? Even with healthy majorities in both houses of Congress, passing cap-and-trade legislation turned out to be a tough sell. In his first address to Congress, he had asked for cap-and-trade legislation to drive the transition to renewable energy. In June that year, the Waxman-Markey bill passed the House of Representatives by seven votes. Just like the EU, it would have created a cap-and-trade regime and imposed a 20 percent renewables target by 2020, which would have destroyed the very carbon market it was creating, just as had happened in Europe. Waxman-Markey didn't make it out of the Senate. The collapse of the Copenhagen climate summit at the end of 2009 dashed hopes for a binding international agreement.* By July 2010, Senate majority leader Harry Reid had given up on cap and trade. Another route had to be found.

The following month Bob Sussman, senior policy counsel to EPA Administrator Lisa Jackson, who had come to the agency from John Podesta's Center for American Progress, met with the Natural Resources Defense Council's (NRDC) top lobbyist, David Doniger. They appear to have decided on a collusive sue-and-settle strategy to legally force the EPA to regulate power-station carbon dioxide emissions under the 1970 Clean Air Act. On August 20, NRDC, the Sierra Club and

* This represents another continuity between President Obama and the last phase of George W. Bush presidency: "And let us complete an international agreement that has the potential to slow, stop, and eventually reverse the growth of greenhouse gases" (State of the Union address, January 28, 2008). For more on the Copenhagen climate conference, see Chapters 31–33 of Rupert Darwall, *The Age of Global Warming—A History* (London, 2013).

the Environmental Defense Fund (EDF)—organizations that would collectively received $96.5 million from the top ten green foundations in the four years to 2013[19]—wrote a joint letter threatening legal action against the EPA.[20]

Recourse to the Clean Air Act was not where the Obama Administration wanted to end up. Writing in 2014 after he'd left the EPA, Sussman says that using the Clean Air Act was a way of strong-arming Congress into passing cap and trade. As a climate tool, it was seen as "poor and probably unworkable."[21] In December, a top White House economic official briefed reporters on the administration's strategy. "If you don't pass this legislation, then . . . the EPA is going to have to regulate in this area," the official—speaking on condition of anonymity—told them.

> And it is not going to be able to regulate on a market-based way, so it's going to have to regulate in a command-and-control way, which will probably generate even more uncertainty.[22]

Alluding to the sue-and-settle strategy at his postelection press conference in November, the President noted that the EPA was under a court order that greenhouse gases are a pollutant falling under its jurisdiction. The "smartest thing" would be for Democrats and Republicans who agreed with him to get in a room to find ways to move forward. Switching from present to past tense, the President implicitly conceded that the legislative path had closed:

> Cap and trade was just one way of skinning the cat; it was not the only way. It was a means, not an end. And I'm going to be looking for other means to address this problem.[23]

Sue-and-settle came at a cost, not least to EPA's integrity as a public agency. It transferred power from the legislative to the executive branches and thence to unaccountable NGOs funded by West Coast billionaires and progressive foundations. Reminiscent of Germany's *Energie Putsch* masterminded by lobbyist/legislator Hermann Scheer ten years earlier, public policy was captured by green ideologues (Chapter 12). The extent of the collusion between the EPA and the NRDC and other NGOs in using the threat of litigation to capture federal policy and how the contrived sue-and-settle stratagem required EPA's full complicity is laid bare in a 2015 report by the new majority staff of the Senate EPW committee under its Republican chairman, James Inhofe.[24] It was something the EPA didn't want the world to know about. Responding to a July 2014 *New York Times* report on the role of NRDC top staff in drawing up the EPA clean power rules, EPA Administrator Gina McCarthy emailed staff,

> If you're laughing right now, it's because you know just how preposterous that is. That is not how our process works. The work we do is too important to the health and the future off [*sic*] of all Americans to be left to the influence of any single outside group.[25]

It was Clintonesque parsing; the collusion involved more than one outside group. Two days before Christmas 2010, EPA announced it had reached a settlement deal with NRDC and the two other NGOs requiring EPA to issue power-station greenhouse gas emission rules before the end of President Obama's first term. NRDC's Doniger emailed McCarthy, then assistant administrator for EPA's Office of Air and Radiation, thanking her for the announcement and promising: "We'll be with you at every step in the year ahead."[26] In words closer to the truth than those to EPA staff and what she told the Senate, McCarthy replied: "This success is yours as much as mine."[27]

In reality, they were on the same side playing on the same team. The green revolving door between NGOs and the EPA under the Obama Administration is one of the features of the Climate Industrial Complex criticized in the 2014 minority staff EPW report. Over a dozen senior EPA positions had been filled by NGO and foundation personnel, and a similar number had moved in the opposite direction.[28] After the 2008 election, the number of NRDC staff hired to draft climate legislation on the Hill and at the EPA led to talk of an NRDC mafia. "We can't equate our revolving door with industry. Theirs involves millions of dollars," an NRDC source told the *New York Times* in March 2009.

> People go from modest means in nonprofits to modest means in government. It's not about the money.[29]

There's money and there's power. Green ideology is a powerful motivator. Oftentimes saving the planet and making money come together. In December 2009, a public affairs VP for Siemens, the German wind-turbine manufacturer, which had spent $5 million on lobbying that year, emailed McCarthy's predecessor, Lisa Jackson, to ask if she would meet Siemens's global sustainability officer: "She'd like to meet you and to express her support for your good work." Jackson agreed. "P.S. Can you use my home email rather than this one when you need to contact me directly? Tx, Lisa," the EPA administrator wrote in a second email. "This reflects a clear intention to violate law and policy," charged the Competitive Enterprise Institute's Christopher Horner, whose Freedom of Information Act requests would reveal Jackson's illicit use of her private email account.[30] Jackson left the agency in February 2013 amid congressional probes into EPA record keeping—there was a five-month gap before McCarthy took over. The revolving door spun, and at the end of May, Apple announced that Jackson was being hired as the company's senior environmental adviser.

The path to EPA regulation had been opened by the Supreme Court in *Massachusetts v. EPA* in April 2007 in what must be counted as the Climate Industrial Complex's single greatest triumph. The case had stemmed from a 1999 suit brought by a group of 19 NGOs (including Friends of the Earth and Greenpeace) and green special interests (including Bio Fuels America and the Solar Energy Industries Association) to have the EPA regulate greenhouse gases in vehicle exhaust. After the Bush Administration's EPA reversed its predecessor's position and decided that greenhouse gases were not air pollutants as defined in the Clean Air Act, Massachusetts and 11 other states and 13 NGOs (NRDC, EDF, the Sierra Club, and the Union of Concerned Scientists joining Friends of the Earth and Greenpeace) sought review of the EPA. By five to four, the Court decided that carbon dioxide fit well within what it called the Clean Air Act's "capacious definition" of air pollutant. That didn't make carbon dioxide a pollutant or force EPA to regulate it. According to Justice Antonin Scalia, writing for the Court in the 2014 case *Utility Air Regulatory Group v. EPA*, which the agency lost, the definition was a description of the universe of substances EPA "may *consider* regulating" under the act's operative provisions.[31] Were EPA to make an endangerment finding, "the Clean Air Act requires the agency to regulate emissions of the deleterious pollutant from new motor vehicles," the Court ruled in *Massachusetts v. EPA*.[32]

The Obama EPA duly produced one to coincide with the Copenhagen climate conference—and had to confront a similar problem that had faced the plaintiffs in *Massachusetts v. EPA*—namely, how to demonstrate that global warming would be harmful to Americans. Massachusetts had alleged loss of its coastline. Remediation costs alone could run well into the hundreds of millions of dollars, it was asserted.[33] However, the details dissolved when viewed through a magnifying glass. Projected sea-level rise of 20 to 70 centimeters by the end of the century was similar to the average error of 30 centimeters and maximum observed error of 70 centimeters in mapping coastline elevation. As Chief Justice John Roberts pointed out in his dissent, if the elevation of coastal land is underrepresented to an extent equal to or in excess of the projected sea-level rise, "it is difficult to put much stock in the predicted loss of land."*[34]

Similar sounds of barrels being scraped can be heard in the EPA endangerment finding. The EPA expected emissions of "well mixed greenhouse gases"—the finding was not in respect of carbon dioxide alone—to lead to higher levels of ozone. Higher temperatures might cause a net increase in morbidity and mortality—or they might not. There was potential for higher crop yields but also a risk of an adverse effect on crop markets. Higher carbon dioxide levels might lead to more hay fever. Climate change would mean more electricity being produced at peak times. Despite the weak observational basis for making the claim, the EPA thought its best shot was the risk of more, and more intense, weather events,† an eventuality that, the EPA said, "clearly supports a finding of endangerment."[35] As we shall see in Chapter 21, it's not straightforward to make the numbers show that global warming presents a dire threat to the welfare of Americans.

Here again, NRDC help was on hand. In June 2011, McCarthy and Doniger met at a Starbucks a block from the EPA's headquarters. The

* The Supreme Court also broke legal ground in establishing Massachusetts's standing. In foregoing its sovereignty in becoming part of the United States, the state had given up its ability to break international law as it "cannot invade Rhode Island to force reductions in greenhouse gas emissions" (p. 16).

† According to a 2012 IPCC special report on extreme events, increasing exposure of people and economic assets was the major cause of long-term increases in economic losses from weather- and climate-related disasters. "Long-term trends in economic disaster losses adjusted for wealth and population increases have not been attributed to climate change, but a role for climate change has not been excluded (high agreement, medium evidence)." IPCC, *Managing The Risks of Extreme Events and Disasters to Advance Climate Change Adaptation*, (2012), p. 9.

next day, the NRDC's Dan Lashof sent the EPA's head of policy and fellow NRDC mafia member Michael Goo its 2011–2012 climate advocacy strategy, subtitled "Briefing for Administrator Jackson." Included in the deck was a section on the lessons from California and the defeat of Prop. 23. Polling showed that only 7 percent of people surveyed had said that climate legislation was important to counter global warming, and only 6 percent said it was needed to protect the environment. Air pollution/health risks were shown to be the most powerful factor supporting climate policies. The NRDC therefore recommended messaging around "people not polluters"[36] drawn straight from the progressive playbook of the People vs. Polluting Corporations. So, like acid rain before it, the justification for regulating power-station emissions was shifted from damage to the environment to harming public health.

As with sue-and-settle, the EPA had a posse of outside helpers to collude with in mounting the public health case against carbon dioxide. Judged by the amount of EPA funding it gets, the American Lung Association (ALA) is the EPA's most valued public health messenger. Between 2009 and 2014, the ALA received $13.97 million from the EPA, making it the top environmental nonprofit recipient of EPA funding.[37] According to George Mason University economist James Bennett, the ALA, originally formed in 1892 to prevent tuberculosis, acts as a "public relations flack" for government agencies, becoming a powerful lobbying organization by selling its reputation.[38] In return for EPA funding, the ALA gave EPA's climate propaganda a veneer of medical plausibility, putting out a stream of press releases supporting the agency's positions. After the defeat in November 2011 of Senator Rand Paul's attempt to rein in the EPA, the ALA praised the bipartisan majority for standing up for the health of our children against "big corporate polluters."[39] Straight out of the progressive playbook, the press release could have been written by the NRDC. The ALA also litigated the EPA at the behest of the EPA. By one count, the ALA was involved in 32 sue-and-settle suits with the EPA.[40]

EPA's first order of business after the endangerment finding was to regulate auto emissions. Released in July 2011, President Obama declared the EPA's fuel economy proposals "the single most important step we've ever taken as a nation to reduce our dependence on foreign oil."[41] A year later, the White House was boasting that the consumer savings from the finalized rules were comparable to lowering the price of gasoline by one dollar a gallon by 2025.[42] It sounded like a big deal. Evidently the White House had forgotten about fracking. By the end of 2015, the price of gas had fallen by $1.50 per gallon. The EPA's regulatory impact assessment assumed a gas price of $3.87 per gallon in 2025 compared to just over $2 per gallon at the end of 2015. The EPA reckoned that the new fuel-economy rules would save consumers $212 billion in 2050.[43] Even if those estimates turn out to be right, it was steam-age progress in an age of an IT-driven energy transformation. Consumers didn't have to wait 35 years. The fall in gas prices to the end of 2015 meant consumers would already be saving $210 billion a year.[44] American capitalism was handily beating environmental regulators at delivering actual cost savings and providing a timely reminder that when the price of hydrocarbons goes down, the cost of decarbonization goes up. As Mark Mills and coauthor Peter Huber write in *The Bottomless Well* (2005), the economics of efficiency depend on the cost of what efficiency saves.[45]

Indeed, fracking was fast becoming greens' worst nightmare. It had killed the national security argument against fossil fuels. In 2013, America surpassed Saudi Arabia in combined oil and natural gas liquids output and surpassed Russia in natural gas.[46] It destroyed the argument for renewable energy based on the specter of looming resource shortages. Hydrocarbons were as abundant as they'd ever been. Reason should have laid to rest the case for renewable energy, but rationality was a casualty of green ideology and the insatiable greed of green capital. For the Climate Industrial Complex, the fuel-economy rules were the appetizer. Regulating power station emissions was the entrée.

Not for want of trying on the part of the EPA, by summer 2011 the original timetable in the sue-and-settle agreement had slipped badly. In September the heads of 19 NGOs wrote President Obama a peremptory letter about the delay and demanded that the EPA stick to the agreed remedial timetable and issue the proposed power plant rules as soon as possible in 2012. The NGOs were not writing to the President as supplicants but as parties to a legal settlement. As Doniger explained, the litigants were not prepared to wait until the President's "hypothetical second term" and have the rules kicked over into 2013.[47] As it turned out, the complexity of using the Clean Air Act defied the EPA's efforts to issue them before the end of 2012.

In his first State of the Union address after his reelection, the President told Congress that if it didn't act, "I will direct my Cabinet to come up with executive actions we can take." He went on to give one of the strangest justifications for renewable energy yet. "As long as countries like China keep going all in on clean energy, so must we."[48] Why? Wasn't America different? China's solar industry had been brought into being by German feed-in tariffs that had saddled Germany with electricity prices second only to wind-dependent Denmark (Chapter 13). Only America had the technology, the people, and the economic dynamism to pioneer and rapidly scale fracking. And the President of the United States wanted America to follow China. It was a new twist to the Obama doctrine of leading from behind.

21
"Our Kids' Health"

> The great thing about this proposal is that it really is an investment opportunity. This is not about pollution control. . . . It's about investment in renewables and clean energy.
>
> *EPA Administrator Gina McCarthy, July 2014*[1]

> Right now, our power plants are the source of about a third of America's carbon pollution. . . . For the sake of our kids and the health and safety of all Americans, that has to change,
>
> *Barack Obama, August 2015*[2]

On August 2, 2015, President Obama announced the finalized Clean Power Plan rule. It was worth the wait. According to that morning's *New York Times*, it would be tougher than earlier versions. The President had come to see the plan in terms similar to the Affordable Care Act in being central to his legacy and had moved to strengthen the rule. Earlier versions would have allowed states to lower emissions by transitioning from coal to gas. The final rule would push electric utilities to bring forward investments in renewables and set off a wind and solar boom, raising wind and solar's share of generating capacity from 22 to 28 percent, the report said.[3]

The projected share of renewable energy would put the United States where Germany had been four years earlier, in 2011.* Germany's renewable energy disaster wasn't mentioned when the President

* At the start of 2011, wind and solar accounted for 26.4 percent of generating capacity, rising to 31.1 percent at year end. *Bundesministerium für Wirtschaft und Energie, Statistisches Bundesamt, Arbeitsgruppe Erneuerbare Energien-Statistik (AGEE-Stat), "Zahlen und Fakten Energiedaten," Stromerzeugungskapazitäten, Bruttostromerzeugung*

launched the Clean Power Plan. Neither was Europe. National security was. The Pentagon was saying that climate change posed "immediate risks" to national security—a risk that hadn't rated a single mention in the 594-page memoir of his first and longest-serving defense secretary. The good news was that over the previous decade, the United States had cut its carbon pollution by more than any other nation, Obama boasted, declining to credit capitalism and the f-word. Then came the bad news. The science was saying that more had to be done. The Clean Power Plan would be, the President said, exactly as the EPA fuel-economy rules had been for reducing dependence on imported oil four years earlier, "the single most important step America has ever taken in the fight against global climate change."[4]

Seven years earlier, the talk was of slowing the rise of the oceans and healing the planet. Now the rhetoric was decidedly downbeat: "We can do more to slow, and maybe even eventually, stop the carbon pollution." There was less talk about the planet and more about the air children breathe and public health. By 2030, premature deaths from power plant emissions would be cut by nearly 90 percent, and there would be 90,000 fewer cases of asthma among children each year, the President declared (applause). Moms across America were spreading the word about how climate change was endangering children's health.

Like a Countrywide mortgage salesman pushing dodgy subprime loans, the President dismissed the plan's objectors. "They will claim this plan will cost you money." The analysis showed it would reduce the average American's energy bill by nearly $85 a year, he countered. "To scare up votes, they'll say the plan is a war on coal." It is. "Even more cynically, they will claim that the plan would harm minority and low-income communities." It does. He wrapped up by recalling past environmental cleanups. Some were genuine: the smogs of Los Angeles (The President recalled, as an 18-year-old, arriving in L.A.: "I wanted to go take a run. And after about five minutes, suddenly I had this weird feeling, I couldn't breathe.") and Ohio's Cuyahoga River—one of the most polluted in the nation—catching fire in the 1960s. The big one cited by the President was not genuine: acid rain threatening

---

und Bruttostromverbrauch Deutschland, Tab 22, http://www.bmwi.de/DE/Themen/Energie/energiedaten.html.

to destroy all the great forests of the Northeast. "And those forests are still there." Had anyone from EPA told the President the truth about acid rain and that the forests would have still been there if the EPA and Congress had done nothing? "So we got to learn lessons," he continued, "We got to know our history." Quite. This was not an occasion for truth telling, especially about the EPA's role in misleading the nation in a previous environmental scare.

Every one of the President's claims was disingenuous at best. Coal is the biggest loser from the Clean Power Plan. Wind and solar energy hits the poorest hardest. It took only three years for Germany's *Energiewende* to increase the number of households trapped in fuel poverty by one-fourth. What about the promised $85 a year off electricity bills? The German Greens had claimed *Energiewende* would cost only a scoop of ice cream a month. Nine years later the cost was pushing one trillion euros ($1.08t). Now President Obama was saying Americans would get the equivalent of seven scoops off their monthly electricity bills.

How could that be? How could the EPA in its regulatory impact analysis (RIA) claim that the Clean Power Plan would result in electricity prices being between 0.01 percent and 0.8 percent higher in 2030 than they would otherwise be, or that the extra cost that year would only be $5.1 billion to $8.4 billion, when the cost of renewable feed-in tariffs in Germany runs at over €20 billion ($22b) a year and America consumes over six times more electricity than Germany?[5]

President Obama and the EPA had not defied the laws of physics to conjure up a miracle of pain-free, near-zero-cost renewable energy. Despite Europe's being further down the path to the renewable nirvana, the 343-page RIA barely refers to the continent's unhappy experience with renewables—a grand total of two paragraphs on the employment implications of electric utilities participating in cap and trade. Instead, the RIA served up estimates purportedly showing a 2030 cost–benefit ratio for the Clean Power Plan of between 3 and 16 times.[6] Worse than the made-up numbers is the manipulation behind them. To arrive at the conclusion that the Clean Power Plan produces huge benefits at virtually no cost, the EPA needed to perform the regulatory equivalent of a three-card trick. It had a flaw. According to the EPA's contrived logic, because the plan doesn't make much difference, it doesn't cost very much. In EPA world, what Obama called America's big step in tackling climate change isn't such a big step after all.

Like every successful con, maintaining audience credulity depends on preventing the audience from noticing what the trickster is up to. The first part is known in the trade as defining the baseline. In regulatory impact assessments, regulators must show what the world would look like without the new regulation. The EPA trick here was to define a business-as-usual baseline that is uncannily similar to the world without the Clean Power Plan. Under its business-as-usual scenario, the EPA asserts that "natural-gas fired generation and renewables would be expected to increase without this rule."[7] For an epoch-defining rule, it would, according to the RIA, have a modest impact on America's generating mix. Although coal is foreseen as the biggest loser, projected to contract by 28.5 percent by 2030 (Table 4, below), the fall under the Clean Power Plan is "consistent with" recent historical trends. Trends for all other types "will remain consistent with what their trends would have been in the absence of this rule."[8]

The renewables trend claim was fabricated to conceal the plan's

Table 4: EPA Clean Power Plan in 2030 (Rate Based) Compared to 2014

	2014	2030 (rate-based)	Change 2014–2030	
	Thousand GWh	Thousand GWh	Thousand GWh	Percent
Coal	1,582	1,131	−451	−28.5%
Natural gas	1,127	1,357	+230	+20.4%
Wind, solar, and other renewables	279	488	+209	+74.8%
Nuclear	797	777	−20	−2.5%
Hydro	253	341	+88	+34.7%
Other	42	28	−14	−33.3%
Total	4,080	4,122	+42	+1.0%

Source: U.S. Energy Information Administration, "January 2016 Monthly Energy Review" (January 27, 2016), Table 7.2a Electricity Net Generation: Total (All Sectors) & EPA, *Regulatory Impact Analysis for the Clean Power Plan Final Rule* (October 23, 2015), Table 3-11.

costs. The year 2012 had been a boom one for onshore wind. In 2013, new onshore wind capacity fell off a cliff, dropping by 92 percent. A bounce in 2014 saw new onshore wind installations climb to less than two-fifths of the 2012 level.[9] Although the 209,000 GWh increase in renewable generation envisaged by the EPA (rate-based scenario) is less than the 230,000 GWh from natural gas, it starts from a much lower base. In percentage terms, the increase in renewable generation—the vast majority being wind power—is over three and a half times greater (74.8 percent to 20.4 percent). Testifying before the Senate Environment and Public Works Committee in July 2015, EPA administrator Gina McCarthy gave the game away. The Clean Power Plan was really an investment opportunity, she told senators. "This is not about pollution control. . . . It's about investment in renewables and clean energy."[10] The trend claim was merely a regulatory contrivance to hide the plan's true costs.

The renewables boom envisaged by the EPA is breathtaking. The Clean Power Plan's underlying best system of emissions reduction projections assume construction of 104,317 MW of new wind capacity between 2022 and 2030.[11] This equates to around 45,000 2.3 MW turbines covering approximately 5.2 million acres—equivalent to over four-fifths of the state of Vermont.[12] The renewable build-out projected for 2013–30 assumes that the United States has the capability to install more wind and solar capacity than the world had in 2012.[13] It assumes that Texas—America's number one wind state, with over 17,000 MW of wind capacity—adds more wind and solar than any nation had in 2012.[14] Yet in 2030, according to the EPA business-as-usual world, total power-sector generating costs would be between $18.0 billion and $21.2 billion higher *without* the Clean Power Plan.[15] A renewables investment boom that doesn't cost consumers a penny? This was beyond even the powers of the EPA in its make-believe world.

Despite President Obama's denial, it's no surprise that coal would be the Clean Power Plan's biggest victim. But the identity of the next-biggest loser constitutes EPA's second RIA trick. Whereas coal is projected to see a 451,000 GWh reduction, electricity consumption is expected to see a 344,000 GWh cut compared to business as usual (Table 5). In 16 years, the Clean Power Plan envisages electricity consumption rising just one percent.

Table 5: Electricity Consumption: EPA Business as Usual Compared to Clean Power Plan (Rate Based)

	2014	2030	Change 2014–2030	
	Thousand GWh	Thousand GWh	Thousand GWh	Percent
Business as usual	4,080	4,466	+386	+9.5%
Clean Power Plan (rate based)	4,080	4,122	+42	+1.0%
Difference		−344		−7.7%

Source: U.S. Energy Information Administration, "January 2016 Monthly Energy Review" (January 27, 2016), Table 7.2a Electricity Net Generation: Total (All Sectors) & EPA, *Regulatory Impact Analysis for the Clean Power Plan Final Rule* (October 23, 2015), Table 3-11.

Demand-reduction cuts the costs of generating electricity (the answer to the EPA riddle of how the Clean Power Plan cuts generating costs by $18.0b to $21.2b). It does so by transferring cost to businesses and consumers, who are expected to invest in energy-saving technology. These don't come cheap. According to the EPA, the annualized costs of these in 2030 are between $26.3 billion and $32.5 billion—a stealth tax on American businesses and consumers that does nothing for the deficit.[16] Demand reduction had originally been one of the plan's four blocks, but as even the EPA had to concede, it fell far outside traditional interpretations of the relevant provision, section 111(d) of the Clean Air Act. Nonetheless, the plan's CO_2 reductions depend on Americans hardly using any more electricity—and that assumes zero population growth.

Even though the Clean Power Plan numbers form the basis of America's international commitment as part of the Paris Climate Agreement to reduce greenhouse gas emissions by 26 to 28 percent below their 2005 level by 2025, there is no existing or planned compliance mechanism to prevent Americans from consuming more electricity. In its Intended Nationally Determined Contribution submitted to the UN, the Obama Administration asserts that "a number of existing laws, regulations and other domestically mandatory measures are relevant to the implementation of the target."[17] As far as the component of these committed reductions from energy efficiency is concerned, it rests on nothing more substantive than hot air.* Meeting the demand reduction target implies almost as many years in which consumption declined as it grew. In the 65 years from 1949 to 2014, there have been only 7 years in which electricity consumption fell—the recession years of 1974 and 1982, the near-recession of 2001, and during the subprime crisis and its aftermath (2008, 2009, 2011, and 2012). The weakness of the Obama economy is reflected in average electricity consumption growth of less than one percent a year from 2009 to 2014.[18] Zero growth in electricity demand requires the economy to remain chronically weak.

Energy efficiency has been part of the green catechism since the first energy crisis of 1973; the cheapest new power station is the one that isn't built. In *The Bottomless Well,* Mark Mills and Peter Huber provide a high-grade demolition of the greens' energy efficiency fantasy. Rising efficiency, they write,

> certainly does not guarantee falling energy consumption. Through all of technological history on record so far, it has had just the opposite effect.[19]

Energy-efficiency policies inevitably fail, they observe: "The more energy efficient a technology grows, the faster it metastasizes and finds new applications."[20] Where environmentalists see energy production and find enormous amounts of it going to waste, Mills and Huber see the Second Law of Thermodynamics. Energy isn't produced (or destroyed). It is made more ordered. What permits order to increase is not the input of high-grade energy but the dumping of low-grade energy (the condensing tower of power stations; the fans expelling heat from a microprocessor) to its surroundings. "A system sheds 'entropy'—chaos—only by shedding energy itself."[21]

In making the argument that curbing the amount of wasted energy means we can consume a bit more without producing any more,

* Under the Clean Power Plan, states could chose between a rate-based option for coal- and gas-fired plants, defined in terms of the maximum pounds of CO_2 per net MWh produced, and a mass-based option, measured in terms of statewide emission of tons of CO_2. The way the EPA converted the rate-based to the mass-based option pushed states to opt for the latter, to encourage development of a nationwide cap-and-trade market. However, the former does not cap total emissions. States opting for the rate-based control would have therefore escaped the targeted CO_2 reductions represented by energy efficiency.

environmentalists flout a fundamental law of physics and ignore the lessons of economic history. Energy consumption rises not solely because economies grow larger. Innovation demands more refined and greater quantities of higher-quality energy. "Most of our demand for energy," Mills and Huber point out

> derives from energy's capacity to refine energy, and from power's capacity to purify power. *Our main use of energy—by far the most important in the "energy racket"—is to purify energy itself.* It is only by grappling with that strange fact that we come to understand why we use so much energy, and why we will always demand still more.[22]

James Watt's steam engine pumped water from coal mines; lasers consume light to generate a beam of highly concentrated, ordered light; power caches on microprocessors ensure a steady flow of electrons switching electronic logic gates operating at billions of cycles a second. Rising energy consumption is the very stuff of progress. Policies to prevent it rising would condemn America to European-style stagnation.

Pot smokers are also blowing a hole in President Obama's carbon caps. "Consumers seeking a green lifestyle are likely unaware that their cannabis use could cancel out their otherwise low-carbon footprint," researcher Evan Mills told Bloomberg's Jennifer Oldham. Indoor cannabis production consumed nine percent of California's household electricity, and some larger growers were spending a million dollars a month on electricity. The intense heat from dozens of 1,000-watt light bulbs requires air conditioning and fans to keep grow rooms at 75 degrees, a dehumidifier to prevent mold, and a carbon dioxide injection system to accelerate plant growth and allow growers to reap multiple harvests a year. Planners feared that escalating marijuana consumption could, in some regions, "undo Americans' attempts to save energy by buying more efficient refrigerators, washers and hair dryers."[23]

Having planed down the Clean Power Plan costs to a bump in the curve of ever-rising wind and solar power costs (trick one) and imposed a stealth energy efficiency tax on American consumers (trick two), the EPA needed to show Americans what they'd be getting in return. Trick three demanded considerable creativity. Global warming is global. A 2013 interagency paper on the social cost of carbon

(SCC) recognized that global warming posed particular analytical challenges in meeting the Office of Management and Budget requirement that economically significant regulations be analysed from a domestic perspective. Emissions in the United States contribute to global damages. The SCC derived from integrated assessment models incorporates the global, not the domestic, cost of damages (now and in the future) from greenhouse gas emissions. Moreover, the United States cannot solve the problem unilaterally. "Even if the United States were to reduce its greenhouse gas emissions to zero, that step would be far from enough to avoid substantial climate change," the interagency group noted.[24]

Numbers from the PAGE09 integrated assessment model, an updated version of one of three such models used by the group, illustrate the relative scale of American contributions to global warming damages and the benefits of climate policies. On the basis of a $60 social cost per short ton of carbon dioxide (one of the four values used in the RIA), $9.54 worth of the $60 of global damages is caused by emissions from the United States. If the United States incurred $60 of costs to reduce its emissions by one ton, only $4.08 of benefits from the $60 would flow back to the United States in the form of avoided damages, implying a domestic cost–benefit ratio of less than 0.1.[25] The model also illustrates the lopsided contributions between industrialized and developing nations' emissions and damages. Less than 10 percent ($5.52) of the $60 of damages from a ton of carbon dioxide impacts industrialized nations (defined as Annex I of the UN framework convention on climate change) from their own emissions, while over 45 percent ($27.24 of the $60) comes from extra impacts in the rest of the world from their own emissions.[26]

The Clean Power Plan RIA airbrushes away the export of 84 percent of climate mitigation benefits from the United States by saying that there was no "bright line" between domestic and global damages. "Adverse impacts on other countries can have spillover effects on the United States, particularly in the areas of national security, international trade, public health and humanitarian concerns"—in other words, the classic rationale for foreign aid.[27] A study by 13 academics cited by the EPA argued in favor of focusing on global rather than domestic damages. In a dog-chasing-its-tail circular argument, they wrote that the United States faced a strategic foreign relations question.

> The United States is engaged in international negotiations in which US emission reductions are part of a deal for abatement by other countries. . . . Our view is that these are compelling reasons to focus on a global SCC.[28]

Even so, an effective global deal would at most raise the proportion of climate benefits retained by the United States from around zero to 16 percent.

In terms of the incidence of its costs and benefits, climate policy is a vast foreign aid program. It is exactly as the Potsdam Institute's Ottmar Edenhofer put it: the de facto redistribution of the world's wealth. Using the one trillion euro cost of the German *Energiewende* as benchmark, assuming that a lower proportion of wind and solar in America halves costs and a one-third cost improvement on top of that, then scaling up for America's higher electricity consumption implies a total cost of $2.3 trillion. Spreading the cost over two and a half decades gives an annualized cost of $92 billion. For fiscal year 2015, the State Department requested $35.6 billion for its various foreign aid programs.[29] Without congressional authorization, America's *Energiewende* implies more than tripling America's foreign aid program. Those numbers are a tough sell.

The EPA therefore needed to show nonclimate benefits from reducing power station emissions. The Obama Administration also needed to bring back an agreement from the Paris climate conference and claim that the developed nations would not be the only ones cutting their greenhouse gas emissions; otherwise the climate rationale for the Clean Power Plan would completely collapse. The EPA resorted to what it had done with acid rain—pitch the benefits of emissions cuts in terms of better public health. Thus, the RIA discovers "air quality health co-benefits" in 2030 of $12 billion to $34 billion compared to climate benefits of $20 billion (both discounted at three percent).[30] The plan could be pitched as a buy one, get one free deal. The small print reveals it wasn't all it was cracked up to be: It breaks the Tinbergen Rule. Named after Jan Tinbergen, the Dutch economist and winner of the 1969 Nobel Prize in economics, the rule stipulates that there must be at least one policy instrument for each policy target. If there are fewer tools than targets, then some targets and therefore some goals will not be achieved.[31] Or, as we say in Britain, horses for courses.

Reflecting the polling data the NRDC had given EPA four years earlier, the RIA devotes fewer than 10 pages to the climate benefits and 46 to health and other co-benefits. Improved visibility from the reduction of haze even makes a reappearance, this time as an unquantified co-benefit. As with acid rain, the bulk of the air quality co-benefits (78–86 percent) arises from reductions in sulfur dioxide emissions, chiefly from reducing the concentration of very small airborne particulate matter, $PM_{2.5}$—microscopic particles measuring 2.5 micrometers and less. There are numerous ways $PM_{2.5}$ gets into the air. Formation of $PM_{2.5}$ from sulfur dioxide is one of them. This makes sulfur dioxide a $PM_{2.5}$ precursor gas, although, as a report posted on the EPA website acknowledges, there is a general consensus that the formation of $PM_{2.5}$ from precursor compounds is "highly uncertain."[32] Insofar as $PM_{2.5}$ presents a threat to public health, it can be said with certainty that by breaking the Tinbergen Rule, a plan that is all about cutting carbon dioxide emissions is the wrong way to tackle it. Trees, for example, remove fine particles from the air. A 2013 study of ten major U.S. cities found that the value of $PM_{2.5}$ removed from the air by trees ranged from $1.1 million a year in Syracuse to $60.1 million a year in New York.[33]

Unlike greenhouse gases where the effect is global, the RIA's valuation from cutting sulfur dioxide emissions is based on the geographic distribution of the EPA-modeled proposal. It warns that these "may not reflect local variability in population density, meteorology, exposure, baseline health incidence rates, or other local factors for any specific location." The resulting estimates, the EPA says, "should be interpreted with caution."[34] In addition to the pervasive uncertainty of implementing the Clean Power Plan on the effects on $PM_{2.5}$ levels in specific towns and cities, there is lack of knowledge about the epidemiology of $PM_{2.5}$ pollutants. Particulates were one of the five pollutants analysed in the 2009 two-decadal study across 11 Canadian cities that found an insignificant or negative relationship between pollution levels and hospital admissions for lung diagnoses (Chapter 7). Furthermore, the Clean Power Plan RIA assumes that, regardless of differences in their chemical composition, all fine particles are equally potent in causing premature mortality.

Uncertainty is thus layered on lack of knowledge on top of conjecture, the most important being the assumption that there is no threshold below which $PM_{2.5}$ ceases to have negative impacts on health down to vanishingly low concentrations. For the EPA, the RIA's beauty is in its vastly multiplying the coincidental benefits it can ascribe to virtually any regulation on virtually anything emitted from combustion. For its mercury rule, 0.004 percent of the health benefits are estimated to come from reducing mercury, and 99.996 percent of the alleged $140 billion health benefits are derived from co-benefits of reducing $PM_{2.5}$.[35] All the experts in a 2010 EPA expert solicitation believed that individuals exhibit thresholds for PM-related mortality. While some disputed the theoretical or empirical existence of a population-level threshold, others favored epidemiological studies as the ideal means of addressing the population threshold issue, observing that "definitive studies addressing thresholds would be difficult or impossible to conduct."[36] Compared to the link between smoking and lung cancer (a relationship that was picked up in the data in the 2009 Canadian study), the scientific understanding of the link between $PM_{2.5}$ and public health (which wasn't) is hardly scientifically robust but gives the EPA enormous scope for regulatory abuse. A 2011 paper by economic consultancy NERA accused the EPA of using its $PM_{2.5}$ co-benefits habit to avoid making the case that other pollutants required tighter controls, charging that

> a high degree of complacency and analytical laziness has instead taken root. . . . The situation is completely at odds with the purpose of RIAs, which is to provide a consistent, credible and thoughtful evaluation of the societal value gained with increased regulatory burden that new rulemakings create.[37]

The no-threshold assumption is regulator's dream. As Ross McKitrick, an author of the Canadian study, explains,

> The problem with the "no-threshold" assumption is that it implies a "never enough" stance on air pollution regulations. No matter how costly and burdensome the rules get, no matter how

> clean the air gets, and no matter how minuscule the potential benefits of new regulations become, EPA bureaucrats get to start each day as if nothing had been accomplished hitherto and the nation is desperately awaiting their help.[38]

No-threshold $PM_{2.5}$ means no risk is too slight and no cost too high. The boundaries are so wide the EPA could have come up with virtually any value it chose. And that is pretty much what it did.

22

Saving the Planet

The absence of evidence isn't evidence of absence.

Attributed to Martin Rees, former Astronomer Royal (on the existence of aliens)

Instead of requiring, for instance, that climatologists prove the reality of global warming, people ask them to prove that cataclysm will not occur. The skepticism that up to that point was seen as an index of wisdom has become a symptom of blindness.

Pascal Bruckner[1]

Like two Gothic columns, the Clean Power Plan and a global climate agreement depend on each other to remain standing. Without one, the other would collapse. Yet their constitutional and legal foundation was as solid as a peat bog. The contrast with acid rain is telling. In 1990 the Clean Air Act Amendments were passed 401 to 25 by the House of Representatives and 89 to 11 by the Senate. Putting scrubbers in power station chimneys is trivial compared to decarbonizing the whole of the power sector, which the Clean Power Plan aims to accomplish by administrative fiat.

The whole edifice teeters on 100 words in the original Clean Air Act. Congress had sought to protect power station owners from the risk of regulatory expropriation. Section 111(d) provides a definition of performance standard as the "best system of emission reduction," taking into account the cost of achieving such reduction in parentheses. "System of emission reduction" had always been defined as a modification to a power station inside the fence and meant that a regulation couldn't set a standard that resulted in the shuttering of the power station because it would have been technically beyond the best system of emission reduction (taking cost into consideration). To operationalize the Clean Power Plan, the EPA overthrew the established interpretation to redefine "system" as embracing the whole grid. A wind or solar farm or a gas-fired power station can then set a performance standard a coal plant cannot meet. Despite its questionable legality, creating facts on the ground would, its supporters thought, make the Clean Power Plan irreversible. Joining the December 2015 Paris climate agreement, structured to avoid a two-thirds vote in the Senate, would create too many obstacles for a future president to repudiate.

European leaders had long been signed up to global warming catastrophism. The annual cycle of UN climate conferences routinely unleashed presentiments of the end of times, especially when they were meant to produce a treaty (Copenhagen 2009) or even a road map to one (Bali 2007; Durban 2011). As the future had once been in Christianity and communism, it becomes again "the great category of blackmail," French philosopher Pascal Bruckner writes in *The Fanaticism of the Apocalypse*.[2] Fear of global environmental catastrophe, Bruckner suggests, is more than a temporary failure of nerve. It is the Old World experiencing a pandemic of weariness.

> Europeans have never been so concerned about the future as they have been since they stopped believing in it. They mention it in order to keep it at bay and to confine themselves to the anxiety regarding the present.[3]

This might be considered frivolous for a continent in decline; it isn't when embraced by the leader of the free world. "ISIS is not an existential threat to the United States," President Obama told *The Atlantic*'s Jeffrey Goldberg. "Climate change is a potential existential threat to the entire world if we don't do something about it."[4] Nothing in the EPA's own endangerment finding warranted such catastrophism. Nor was there anything in the Clean Power Plan RIA, pumped up as it was with fictitious health cobenefits; nor was the commonwealth of Massachusetts so threatened by loss of coastline that would have justified a sovereign Massachusetts invading Rhode Island, as the Supreme Court had mused.

The Clean Power Plan forms the centerpiece of America's nationally determined contribution under the Paris Climate Agreement: "The United States intends to achieve an economy-wide target of reducing its greenhouse gas emissions by 26–28 percent below its 2005 level in 2025 and to make best efforts to reduce its emissions by 28 percent."[5] The requirement to submit a plan does not break new ground. Under the 1992 UN framework convention on climate change, approved by the Senate in a 95 to 0 vote, all parties shall "formulate, implement, publish, and regularly update national programs to mitigate climate change."[6] Where Paris goes further is in imposing an expansive, legally binding ratchet obligation on the United States. "The efforts of all Parties will represent a progression over time," Article Three states, which is elaborated in the strongest language of the whole document in the next article:

> Each Party's successive nationally determined contribution will represent a progression beyond the Party's then current nationally determined contribution and reflect its highest possible ambition.

Thus the Paris Agreement requires the United States to make the Clean Power Plan just the starting point for ever-larger cuts in its greenhouse gas emissions, what the Obama White House hailed as "ratcheting up ambition over time."[7] Moreover, the decarbonization ratchet continues—there is no end date—irrespective of what other parties are doing. In this respect, the Paris mechanism is markedly more problematic than the 1997 Kyoto Protocol. Kyoto had an initial four-year commitment period at the end of which each party's new commitment could take account of what everyone else was doing. With Paris, once on the ratchet, there's no way off other than repudiating the entire agreement.

Pacta sunt servanda—agreements must be kept—is an essential and long-standing principle of international law. Article 26 of the 1969 Vienna Convention on the Law of Treaties states, "Every treaty in force is binding upon the parties to it and must be performed by them in good faith." According to the Vienna Convention definition of a treaty, the Paris Agreement is a treaty in all but name.*

* Although the United States signed the convention in 1970, the Senate has not given its advice and consent. However, the United States considers "many of the provisions of the Vienna Convention on the Law of Treaties to constitute customary international law on the law of treaties." https://www.state.gov/s/l/treaty/faqs/70139.htm (accessed May 8, 2017).

Treaties have a special significance for the United States. Article II of the Constitution circumscribes the power of the executive to make treaties by stating that the President "shall have the power, by and with the advice and consent of the Senate, to make treaties, provided two-thirds of the Senators present concur." Not all international agreements are Article II treaties. In his writings on the Paris Agreement, legal scholar and Clinton-era climate change coordinator at the State Department Daniel Bodansky points out that from George Washington onward, the United States has concluded "congressional-executive agreements" (authorized by a majority of both houses of Congress); "treaty-executive agreements" (authorized under existing agreements); and "presidential-executive agreements" (based on the President's existing legal authority).*

Bypassing Congress was a key requirement of the Paris Agreement. None of the participants wanted a repeat of the Kyoto Protocol, which the Clinton Administration signed but did not send to the Hill for the Senate's advice and consent because it knew what the outcome would have been. France's foreign minister and conference president Laurent Fabius explained six months before the conference, "We know the politics in the US. Whether we like it or not, if it comes to the Congress, they will refuse."[8] That might have been the parties' intention but is not, as lawyers say, dispositive. Although the Nationally Determined Contribution submitted by the United States relies on existing laws and regulations, an economywide target to reduce emissions by 26 to 28 percent below 2005 levels by 2025 is not itself reflected in U.S. law. The measures adopted by the United States might or might not be sufficient to achieve the target, Bodansky argues.

> Consequently, adoption by the president of a domestic implementation commitment without Senate or congressional approval, like adoption of a binding emissions target, would go beyond existing precedents such as the [2013] Minamata Convention [on Mercury].[9]

* Drawing extensively on Bodansky's analysis, as this chapter does, I wrote a piece for NRO (December 14, 2015) on the Paris Agreement questioning the constitutionality of ratifying it without congressional approval, a conclusion Bodansky emailed me to say he disagreed with.

Adoption without Senate approval also rides roughshod over previous executive branch undertakings. During Senate Foreign Relations Committee hearings on the 1992 UN framework convention on climate change, the Bush Administration pledged to submit future protocols under the convention for its advice and consent. When the Foreign Relations Committee reported, it memorialized the executive branch's commitment:

> A decision by the Conference of the Parties [to the convention] to adopt targets and timetables would have to be submitted to the Senate for its advice and consent before the United States could deposit its instruments of ratification for such an agreement.[10]

The question of whether the Paris Agreement is legally binding, Bodansky suggests, ultimately reflects the state of mind, most importantly, of the executive branch officials and judges who apply and interpret the law, what the Oxford legal philosopher H.L.A. Hart called their "internal point of view."[11] A law gives rise to an obligation that there are reasons for obeying and enforcing the law. These reasons are extralegal; they lie outside the system of law itself. Climate change is, Bruckner observes, the great category of blackmail. The widespread belief in global warming catastrophism among elite opinion molds Hart's internal point of view, one shared among the majority of the Supreme Court in *Massachusetts v. EPA*. "Everybody else is taking climate change really seriously," President Obama said at a press conference in Paris on December 1, 2015. "They think it's a really big problem. . . . So whoever is the next president of the United States . . . is going to need to think this is really important."[12] That goes for judges who might harbor doubts about the legality and constitutionality of American ratification of the Paris Agreement or of the Clean Power Plan. As legal challenges to the Clean Power Plan wended their way through the courts, a February 2016 memorandum by the solicitor general to the Supreme Court laid it on the line. Granting a stay to implementation of the plan, Donald Verrilli warned the Supreme Court justices, would harm the public's interest in "preventing the risk of 'catastrophic harm.'" Delaying the rule's implementation, Verrilli continued, "would also disrupt the United States' leadership on the international stage."[13]

The EPA was also adept at creating new facts on the ground by operating at the outer limits of the law and beyond. The agency snubbed its nose at the Supreme Court after a June 2015 ruling against its MATS rule on reducing mercury emissions from coal-fired power stations. The rule had been issued over three years previously; "investments have been made and most plants are well on their way to compliance," it bragged.[14] In December 2015, the General Accountability Office found that the EPA had violated federal law by engaging in covert propaganda supporting its own proposed rules.

In a ruling the previous year, the Supreme Court had struck closer at the legal basis of the Clean Power Plan. In *Utility Air Regulatory Group v. EPA*, the Supreme Court declared the EPA's interpretation of the Clean Air Act unreasonable because it would bring about an enormous and transformative expansion in the EPA's regulatory authority without clear congressional authorization. The ubiquity and scale of carbon dioxide emissions would mean an explosion in the number of businesses requiring burdensome permitting far beyond the EPA's target of electric utilities. To keep annual permit applications from jumping from 800 to nearly 82,000—at one stage, the EPA argued that the number of permits would grow to 6 million—the EPA decided to rewrite the precise numerical pollutant thresholds Congress had specified in the Clean Air Act, clear evidence that Congress had not envisaged the Clean Air Act being used to regulate carbon dioxide emissions in the first place. "We expect Congress to speak clearly if it wishes to assign to an agency decisions of vast 'economic and political significance,'" Justice Scalia wrote for the Court.[15]

To most observers, the ruling didn't appear to threaten the Clean Power Plan. On February 9, 2016, the Supreme Court delivered a thunderbolt. By five to four, it granted a stay on the EPA continuing to implement the plan. The White House expressed its disagreement with the Court. The plan was based on "a strong legal and technical foundation," the White House said, a claim at variance with former administration officials who viewed the Clean Air Act route as "poor and probably unworkable." Nonetheless, the administration vowed to "continue to take aggressive steps to make forward progress to reduce carbon emissions."[16]

Four days later, there was a thunderbolt of a different kind. The Supreme Court became a four-four bench with the suddent death of Justice Scalia. Three days after Scalia's death, the President's climate envoy, Todd Stern, told reporters in Brussels that even if the Clean Power Plan were struck down, "Come what may, we are sticking with our plan to join." It was about creating another fact on the ground: Once signed, no Republican president would be able to withstand the diplomatic cost of repudiating it. Withdrawal would give the country "a kind of diplomatic black eye that I think a president of any party would be very loath to do," Stern said.[17] For a senior diplomat to argue that the United States should take on a treaty commitment with question marks over the legality of the means of meeting it was extraordinarily irresponsible and showed a cavalier disregard for the national interest that should transcend partisan politics. But signing and ratifying would serve the supreme moral imperative of the age. Legalistic quibbles and diplomatic niceties should take second place to saving the planet.

EPA administrator Gina McCarthy was also bullish. Does the stay slow down America's transition to a low-carbon future, McCarthy asked in a speech in Houston a week later. "Absolutely not." What, then, was the point of the Clean Power Plan? Less than seven months earlier, the President had called it the single most important step America had ever taken against climate change. Now McCarthy appeared to be suggesting that fracking had rendered the plan superfluous:

> Over the past decade, the US has cut greenhouse gas emissions more than any nation on earth. And we didn't have the Clean Power Plan ten years ago.[18]

The stay would not reverse the course of history set with the Paris Agreement, McCarthy argued.

The Court's decision had been on "the wrong side of history," Robert Redford chimed in. Deploying Bruckner's great category of blackmail against the surviving four justices who'd ordered the stay, the actor and environmental activist wrote in *Time*:

> The Court has been on the wrong side of History before, but never on a topic that directly affects the current lives of seven billion people, and untold billions in the future.[19]

Four Justices v. Seven Billion People—the case of the century.

In the twentieth century, appeals to the inevitability of history have an uncomfortable ring. Nowadays the brittle rhetoric serves to mask the ideological emptiness of the left. As heirs of the Enlightenment, Bruckner, a philosopher from the left who condemns the left for what it has become, argues that European left-wing movements are thrice guilty of betraying the ideals of the Enlightenment: first, by confusing education with entertainment and for its part in the collapse of schooling; second, by abandoning the battle for equality in favor of identity politics and letting the proletariat drift toward the nationalist right; and third, by deserting the idea of progress, thereby robbing the left of its raison d'être.[20]

The American left's championing of identity politics and its capitulation to environmentalism has seen it forsake its historic role of promoting the interests of working people, fueling the rise of Donald Trump. From being the voice of working people, the Democratic Party has become the political arm of the Climate Industrial Complex, tasked with monetizing environmentalism through a raft of government interventions of one kind or another paid for by working people, and aided by a GOP Congress when it approved generous extensions of the two main subsidies for renewables in the December 2015 budget deal.

Nowhere has the capture of the American left by green rent seeking gone further than in California. "Progress comes from well-designed regulatory objectives," Jerry Brown declared at the Paris climate conference, where he and Tom Steyer had taken a posse of green-tech executives. The four-term governor praised environmental regulations for forcing innovation, a policy that had resulted in refiners being fined for not adding nonexistent "advanced biofuels" after the technology flopped, and told a small crowd to "never underestimate the coercive power of the central state in the service of good." A news conference organized by the Californian delegation featured several beneficiaries of the state's coercive powers. Business owners chuckled when asked if their companies would be viable without government support. "For us," K. R. Sridhar, chief executive of the fuel cell company Bloom Energy, said, "it's a feeding bottle and not an addiction bottle."[21] Green tech sure is one big baby.

Hillary Clinton's quest for the Democratic nomination illustrates the extent of Democrats' surrender to environmental interests. After months of hemming and hawing over the Keystone XL pipeline, in September 2015 she finally came out against it, saying it was a distraction from other issues. Six months later, she did the same on fracking. "By the time we get through all of my conditions, I do not think there will be many places in America where fracking will continue to take place," Clinton said in a debate against Bernie Sanders. "My answer is a lot shorter. No. I do not support fracking," Sanders responded.[22] It signaled an extraordinary retreat from reality—the reality of lower energy prices; of all the jobs, incomes, and GDP generated by fracking; of the transformation of America from hydrocarbon importer to exporter; of the geopolitical advantage America has gained over Russian president Vladimir Putin and the downgrading of the importance of the Persian Gulf; and of a prodigious achievement of American capitalism. It was a retreat from the reality that, because of fracking, the United States had cut its carbon dioxide emissions by more than any other nation. Of course, Robert Redford and other green activists didn't line up to denounce Clinton and Sanders for destroying the Clean Power Plan (as shown in Table 4, natural gas is projected to be the single largest primary energy source for electricity in 2030) and putting America's greenhouse gas commitments in the Paris Agreement beyond reach. None of that mattered. Like Germany turning its back on nuclear power and banning fracking, global warming is a pretext, not a cause, for environmentalists to pursue their war on the energy technologies they dislike.

A new front in the carbon war opened up with calls for ExxonMobil to be investigated under antiracketeering laws for supposedly suppressing the truth about global warming. It had been initiated by Rhode Island senator Sheldon Whitehouse in a May 2015 op-ed in the *Washington Post*. At the end of October, Clinton was asked whether the Department of Justice should investigate the company. "Yes, yes, they should," she replied. "There's a lot of evidence that they misled."[23] The following week New York's attorney general issued a subpoena, demanding extensive financial records, emails, and other documents. The heat was on. In March 2016, the federal attorney general, Loretta Lynch, told the Senate Judiciary Committee that the DoJ had requested the FBI investigate what Senator Whitehouse dubbed the climate denial apparatus.

The allegations centered on a claim made by climate activists that, from as early as 1977, Exxon researchers "knew" the truth about global warming and that the company had hidden or falsified what its researchers had discovered. As a matter of logic, the claim, strongly pushed by Naomi Oreskes and Bill McKibben, would also have required the IPCC to be part of the climate denial apparatus and a party to Exxon's alleged fraud. The IPCC was established in 1988 and tasked with providing a comprehensive review of the science of climate change. Two years later, the IPCC produced its first assessment report. If Oreskes and McKibben are to be believed, the IPCC had been up to its neck in suppressing "the truth" about global warming that Exxon had discovered 13 years earlier:

> Because of the many significant uncertainties and inadequacies in the observational climate record, in our knowledge of the causes of natural climatic variability and in current computer models, scientists working in this field cannot at this point in time make the definitive statement: "Yes, we have now seen an enhanced greenhouse effect."[24]

The allegation was baseless and absurd. But that wasn't the point. The oil companies were being made an offer they couldn't refuse. And it was about putting the First Amendment in the dock.

23

Spiral of Silence

Solitude many men have sought, and been reconciled to: but nobody that has the least thought or sense of a man about him, can live in society under the constant dislike and ill opinion of his familiars, and those he converses with.

John Locke, 1690[1]

It is not so much the dread of what an angry public may do that disarms the modern American, as it is sheer inability to stand unmoved in the rush of totally hostile comment, to endure a life perpetually at variance with the conscience and feeling of those about him.

Edward Alsworth Ross, 1901[2]

In August 2014, the Pew Research Center, an offshoot of the Pew Charitable Trusts, published the results of a survey on people's willingness to discuss contentious issues on social media platforms like Facebook and Twitter. "An informed citizenry depends on people's exposure to information on important political issues and on their willingness to discuss these issues with those around them," Pew explained.[3] If people thought friends and followers on social media disagreed with them, they were less likely to share their views, the survey showed. "It has long been established that when people are surrounded by those who are likely to disagree with their opinion, they are more likely to self-censor."[4] These findings confirmed a major insight of pre–internet era communication studies: the tendency of people not to voice their opinions when they sense their view is not widely shared. The report's authors, led by Keith Hampton of Rutgers University, wrote, "This tendency is called the 'spiral of silence.'"[5]

The Spiral of Silence, published in 1984, was written by Elisabeth Noelle-Neumann, West Germany's foremost pollster. There was more to Noelle-Neumann. As the first sentence of her *Times* obituary put it, Noelle-Neumann moved from working as "a Nazi propagandist to become the grande dame of opinion polling in post-war Germany." A cell leader of the Nazi student organization in Munich, she met Hitler at Berchtesgaden. "She found him sympathetic, lively and engaging."[6] Thanks to a scholarship from Joseph Goebbels's Reich Ministry of Public Enlightenment and Propaganda, she went to the University of Missouri to study journalism. Her 1940 doctoral thesis on George Gallup's polling techniques brought her to Goebbels's notice, who gave her a job writing for *Das Reich*. "To reach into the darkness to find the Jew who is hiding behind the *Chicago Daily News* is like sticking your hand into a wasp's nest," she wrote in June 1941.[7] Dismissed a year later, she distanced herself from the Nazi regime, and after the war she and her husband, also an alumnus of Goebbels's propaganda ministry, established the Allensbach Institute. Turned down by the SPD, Allensbach's services were offered to the CDU. She was soon having tea with Konrad Adenauer, West Germany's first chancellor. "She probably took tea and more with all West German Chancellors."[8]

Noelle-Neumann claimed her thinking about the spiral of silence had been triggered by the 1965 German election, though this was far from the whole story. Polls had shown the CDU–CSU coalition running neck and neck with the SPD, while expectations of the outcome shifted dramatically in favor of the CDU–CSU coalition, accurately forecasting the actual result. Others' opinions might influence one's own behavior, Noelle-Neumann hypothesized. When a population is continuously exposed to a persistent and consistent media account of current events on controversial issues, the primary motivation of a person will be to conform, at least outwardly, to avoid discomfort and dissonance. "Over time there is thus a spiralling of opinion change in favour of one set of views," Noelle-Neumann argued.[9]

The intuition that had led her to the spiral of silence lay outside opinion polls. "The fear of isolation seems to be the force that sets the spiral of silence in motion," she wrote.[10] Historians, political philosophers, and other thinkers provided corroboration. Alexis de Tocqueville had written in 1856 that people "dread isolation more than error."[11] The quotations at the head of this chapter appeared in a lecture given by Noelle-Neumann just two months after the 1965 election. People can be on uncomfortable or even dangerous ground when the climate of opinion runs counter to their views. "When people attempt to avoid isolation, they are not responding hyper-sensitively to trivialities; these are existential issues that can involve real hazards," she wrote in *The Spiral of Silence*.[12] It could be proved

> that even when people see plainly that something is wrong, they will keep quiet if public opinion (opinions and behavior that can be exhibited in public without fear of isolation) and, hence, the consensus as to what constitutes good taste and the morally correct opinion speaks against them.[13]

Evidence came from surveys designed to simulate the threat of social isolation. Respondents were asked questions designed to reveal their willingness to engage in a discussion on a contentious topic with a fellow traveler on a train journey.

> One can see how, as the spiral of silence runs its course, the standpoint that it is unconscionable to smoke in the presence of non-smokers can become dominant to the point where it is impossible for a smoker publicly to take the opposite position. . . . What is being expressed here is quite evidently a cumulative effect; step by step, through hostile responses of the environment, one becomes unnerved.[14]

Train journeys had featured in earlier surveys of German public opinion. Neither was it the first time that Noelle-Neumann had written about opinions expressed on train journeys. The Sicherheitsdienst (SD), the Nazi party's internal security service, monitored German public opinion and devised innovative methods to overcome Germans' fear of speaking frankly to strangers. These included sending trained interviewers on long train trips. The surveys were treated as highly sensitive and kept within an extremely tight circle. There are no known links between Noelle-Neumann and the SD, Christopher Simpson, professor of journalism at Washington, D.C.'s American University, wrote in a 1996 paper, but at least five of her articles in *Das Reich* derived entirely or in part from anonymous interviews on train rides across Germany and coincided with the SD's top-secret train carriage interviews.[15]

Simpson used this implied connection to discredit the idea of the spiral of silence, but as the 2014 Pew report shows, the model is generalizable beyond totalitarian regimes. "My scholarly work was indeed influenced by the trauma of my youth," Noelle-Neumann responded.[16] Indeed, her painfully vivid description of the spiral of silence could hardly have been written by a person who had not lived and worked in a totalitarian regime:

> Climate totally surrounds the individual from the outside; he cannot escape from it. Yet it is simultaneously the strongest influence on our sense of well-being. The spiral of silence is a reaction to changes in the climate of opinion.[17]

After all, Noelle-Neumann worked for a man widely acknowledged as the greatest propagandist of all time. In short, her experience in the Nazi period led her to a truth about the human condition. *Public Opinion—Our Social Skin* is the subtitle of *The Spiral of Silence.*

Niccolò Machiavelli, John Locke, and David Hume had featured in her 1940 dissertation. James Madison was added to *The Spiral of Silence*: "The reason of man, like man himself is timid and cautious, when left alone; and acquires firmness and confidence, in proportion to the number with which it is associated."[18] She argued that Madison would have shared her characterization of public opinion as a fearsome tribunal.*

> Because only the menace, the individual's fear of finding himself alone, as Madison so emphatically described it, can also explain the symptomatic silence we found in the train test and in other investigations, the silence that is so influential in the building of public opinion.[19]

The articulation function of the media helped explain why people holding a majority opinion express unwillingness to engage in conversation in the train test—what one might call the silence of the Silent Majority.

> The media provide people with the words and phrases they can use to defend a point of view. If people find no current,

* In *Federalist* No. 49, Madison argued against Jefferson's proposal to facilitate constitutional change, which, in Madison's eyes, ran the danger of exciting popular passions that a durable constitution needed to withstand.

> frequently repeated expressions for their point of view, they lapse into silence; they become effectively mute.[20]

Media opinion, which reflected the views of a small section of the social elite, was not the same as public opinion. In the long run, the observed majority opinion would not change media coverage, but media coverage would change the observed majority, a process that could happen over weeks, months, and years. The problem, as Noelle-Neumann saw it, was not media manipulation as such, "for journalists only reported what they *saw*."[21] The antidote to the risk of systemic media bias was ensuring a diversity of views: "The apparent consensus arising out of a one-sided media reality can only be avoided if reporters of various political persuasions present their points of view to the public."[22]

One-sided media reporting is a striking feature of the climate and energy debate. "Climate denier" and "tool of malign fossil fuel interests" are epithets used to delegitimize dissent and quash diversity of opinion. "Climate change is a fact," President Obama declared in his 2014 State of the Union address.[23] As philosopher Stephen Hicks argues in *Explaining Post-Modernism*, the postmodern left uses language primarily as a weapon to silence opposing voices, not as an attempt to describe reality.[24] To close the debate down, science masquerading as impartial judge is deployed as lead prosecutor. Dissenters and skeptics are derided as Flat Earthers and scientific ignoramuses. Yet the most stupid utterance on the science of global warming goes without a breath of criticism from scientists who regularly furnish the media with hostile quotes on skeptics' views. "This is simple. Kids at the earliest age can understand this," Secretary of State John Kerry told an audience in Indonesia in 2014. For someone who confessed that he'd found high school physics and chemistry a challenge, climate science, Kerry declared, was easy. The science was "absolutely certain."

> It's something that we understand with absolute assurance of the veracity of that science. No one disputes some of the facts about it. Let me give you an example. When an apple separates from a tree, it falls to the ground.[25]

Fact conflated with theory; certainty where there is pervasive uncertainty and lack of understanding; simplicity where there is unfathomable complexity; climate model predictions of warming elevated above observations. The biggest distortion of climate science is unscientific in its premise and authoritarian in its consequence: "The science is settled. We must act." When systemic media bias is purposed as a tool of state manipulation and social control, a democracy extinguishes its democratic culture. This is not some hypothetical, abstract danger.

By now it should come as no surprise that among the countries that count themselves as formal democracies, Sweden has developed, refined, and deployed mechanisms of social control more comprehensively than any other. German poet and novelist Hans Magnus Enzensberger observed in 1990 that the Swedish state had regulated "the affairs of individuals to a degree unparalleled in other free societies." It had not only gradually eroded its citizens' rights but crushed their spirit.

> It really looked as if the Social Democrats . . . had succeeded in taming the human animal where other different regimes, from theocracy to Bolshevism, had failed.[26]

In his 2014 book *The Almost Nearly Perfect People: The Truth about the Nordic Miracle*, British travel writer Michael Booth was struck by the extent of media censorship in Sweden. Spooked by the rise of the anti-immigrant Danish People's Party, the press in Sweden refused to report the anti-immigrant arguments of the Sweden Democrats. "I think in Sweden we have seen [what's happened] in Denmark. We have seen inclusiveness doesn't work, and we have decided not to do that," journalist and Olof Palme biographer Henrik Berggren told Booth.[27] "The Danish point of view on freedom of speech is quite ridiculous," Stefan Johnsson, a journalist-turned-professor of ethnic studies, commented.[28] (Awarding sinecures in academia is one way the Deep State rewards compliant journalists). Was Sweden totalitarian? Booth asked Sweden's best-known journalist and lead columnist of the *Expressen* tabloid. Ulf Nilson's background was a paragon of social-class compliance. "Your father was a stonecutter," Palme would tell Nilson enviously. "In a sense it is totalitarian," Nilson answered.

> Of course, you can't compare it to Nazi Germany or North Korea, it isn't as bad as that, but there is a creeping totalitarianism which is defined as conforming, to do like others. Nobody really

> questions the kind of society we have, that's what I dislike so much about Sweden. Indoctrination is what you would call it.[29]

Intimidation too. In 2015, Nilson was sacked and branded a racist.

Deep State intimidation and a media blackout did not prevent the Sweden Democrats from becoming Sweden's third-largest party, but in a postelection stitch-up, the center-right opposition parties agreed to support the new Red–Green government as a loyal opposition, the parties switching roles after the 2018 election. The Sweden Democrats weren't the only ones who'd been locked out until 2022. Sweden's voters had been as well.

Sweden's short-circuiting of democracy produced hardly a ripple—testament, perhaps, to the success of a schools system that, as related to Roland Huntford in *The New Totalitarians*, promoted the concept of the "freedom to give up freedom" (Chapter 6). The abstraction of the loss of freedom was becoming a reality. As G. K. Chesterton wrote, despotism might almost be defined as a society grown tired of democracy.

24 The American Republic

The corruption of every government almost always starts with that of its principles.

Montesquieu[1]

America's climate change spiral of silence is a private sector-government coproduction spanning the entire Climate Industrial Complex—universities, multibillion-dollar foundations, Silicon Valley oligarchs, NGOs, EPA bureaucrats, law enforcement agencies, elected politicians, and their helpers in the media. Research for the hit job on ExxonMobil was funded by a number of progressive foundations, including the Rockefeller Family Fund and the Rockefeller Brothers Fund.[2] It was then launched into the public domain by Bill McKibben and his 350.org, also heavily backed by foundation money, and Harvard's Naomi Oreskes. In September 2015, 20 climate scientists wrote to President Obama backing Sheldon Whitehouse's push for corporations and "other organizations" to be investigated under antiracketeering laws. The letter had been organized by Jagadish Shukla, of George Mason University, whose Institute of Global Environment and Society had received more than $25 million in taxpayer grants, its call being taken up by the attorney general of New York and the Department of Justice, subsequently joined by former vice president Al Gore and 16 more state attorneys general, one of whom launched a probe into the Competitive Enterprise Institute, a free-market think tank.[3]

The silencing power of the spiral is greatly amplified by the civil war within American capitalism pitching tech and hydrocarbon companies against each other. This war is not market competition. The two groups do not compete with each other. To varying degrees, they depend on each other; Silicon Valley's data centers would go dark and

Google's fleet of private jets would be permanently grounded without hydrocarbons. It is war because one side denies the right of the other to exist, that the activity of turning hydrocarbons into useful energy is fundamentally illegitimate. And, like the war that was started with the attack on Fort Sumter, it is a war against the idea of America first proclaimed by America's Founders.

In an April 2016 brief filed in the U.S. Court of Appeals by Amazon, Apple, Microsoft, and Google, the self-styled Tech Amici, argued that the Clean Power Plan would be beneficial not only for the environment, it would be good for business.[4] The 32-page brief was replete with claims that the Tech Four knew to be tendentious and misleading. Renewables were reliable, it asserted, a claim flatly contradicted by Google's own engineers, who, just over a year earlier, had concluded that most current renewable technology couldn't provide power that is both distributed and dispatchable. Likewise, Bill Gates had raised a red flag about the inherent unreliability of weather-dependent renewables (Chapter 14). Renewable energy prices were less volatile, the brief asserted, ignoring the reality that output of weather-dependent renewables is unresponsive to demand, making wholesale electricity prices far more volatile, frequently pushing them into negative territory.[5] Only once, in a footnote, did the Tech Four acknowledge the reliability problem of wind and solar. A new Apple campus would, it was claimed, operate with 100 percent renewables, but Apple was working with the local utility to "address the impacts of potential renewable generation intermittency," a dishonest formulation in which "potential" relates to a property inherent in the technology.[6]

Though Schumpeter had predicted in his writings of the 1940s the ultimate demise of capitalism, he not foreseen a war within capitalism as a contributing cause. The Tech Four describe themselves as among the most successful and innovative businesses in the United States. For all that, tech's imprint on the American economy is difficult to discern from macroeconomic data: There is no positive correlation between rising internet usage and productivity growth. On the other hand, tech has generated stupendous wealth for the barons who run companies that aggressively exploit the dematerialized nature of much of their output to minimize their corporate tax bills. In 2015, two of the Tech Four were found to have operated an illegal antipoaching ring designed to suppress employee wages. Along with Adobe and Intel, Google and Apple were ordered to pay $415 million to settle a class-action suit. By contrast with the feudalism of tech, fracking is democratic, spreading wealth more widely and evenly through job creation and higher incomes and generating massive consumer gains by reducing the cost of everything derived from hydrocarbons. Perhaps here we can discern a motive. Ostentatiously donning the cloak of green virtue diverts attention away from tech barons' being targeted as the twenty-first-century's robber barons. Seen this way, tech support for the Clean Power Plan *is* good business: a way of protecting the tech barons' billions.

It is, however, unambiguously bad for the business of America. Europe's *Energiewende* is turning out to be a colossal engine of economic and social disaster. Damaging as they are for America, the implications of the Clean Power Plan go far beyond its economic consequences. How the Clean Power Plan and the Paris Agreement were done matters even more. It shows how Americans are governed now and how they are to be governed in the future. The June 2009 Essen conference on climate change as cultural change was explicit: Tackling climate change will bring about an economic, social, and cultural transformation comparable in scope to the Neolithic and Industrial Revolutions. The transformation they seek would not spontaneously evolve like the previous two but would have to be centrally guided and controlled.

The United States provides the Climate Industrial Complex with firepower that it can get from no other country. Ever since the Rockefeller and other foundations funded the Frankfurt School during its American exile and paid for its return to West Germany, American foundations have been prime funders of environmental alarm, most egregiously when they financed the Soviet disinformation campaign promoting the nuclear winter scare of the mid-1980s. More recently, the Climate Industrial Complex has benefited from the reputational halo and money from the phenomenal success of the American IT industry after Silicon Valley initiated its self-serving civil war against the American hydrocarbon economy.

Yet as participants from both sides of the Atlantic at the Essen conference recognized, the United States is the biggest political obstacle to transforming society through deep decarbonization. America *is* different from every other nation, something many non-Americans admire and many others deeply resent. Whatever they might want to do about it, American exceptionalism will only come to an end at the hands of Americans themselves. There are plenty of those who wish it, for America is a house divided. Powerful domestic forces are pushing the United States to be fully involved in and lead this global transformation. Such a transformation and the means to achieve it are not compatible with the idea of freedom for which Americans declared themselves independent in 1776 and enacted their Constitution in Philadelphia in 1787 to protect. Safeguards to protect freedom were built into that Constitution in a construction of balance and equilibrium that is set out in the greatest document of the Age of Enlightenment.

The eighteenth century had given birth to what, by century's end, the German-born American historian of the Enlightenment Peter Gay calls "the still youthful, always precarious, science of freedom."[7] Other than Great Britain, the Dutch Republic, and the partial exception of Poland, the political tide had been running toward absolutism. In some of the German states and the Hapsburg Empire, the Cameralists—in Gay's words, a tribe of authoritarian rationalists—proposed that the sole purpose of government was the diffusion of happiness, a definition that did not include the freedom of the subject. The people had other duties: "As the ruler must make his people happy the subject, must, by God, obey and *be* happy."[8] It was a vision of government that had absolute monarchy as its most plausible corollary.

> To do his duty adequately, the ruler must have at his disposal a perfectly obedient bureaucracy, all the knowledge it is possible for him to gather, and unlimited authority to translate his programs into law.[9]

Gay's description of absolute monarchy replicates Huntford's description of Sweden's political system that we saw in Chapter 1: a weak legislature and the swift enactment of the intentions of the central bureaucracy, with power residing in the government administrative machine. It is also a good description of President Obama's and the EPA's attitude to the exercise of executive power and the executive's subordination of the legislature, one that is alien to the precepts of that great document of 1787. Modern Sweden is a perfected version of a model of governance that America's Founders rejected when they declared America's independence.

America's uniqueness lies not in the fact of its independence but that it became independent to create something without precedent: government dedicated to the principle of liberty. Continental Europe had proved infertile for the seed of freedom. The Cameralists' ideas were put into practice by several European monarchs. Gustavus III of Sweden was one of six listed by Gay who were branded Enlightened Despots by nineteenth-century historians. Enlightened despotism was the preferred model of governance for progressive European monarchs.

It fell to Americans to establish what Europeans had not—government of the people, by the people, for the people. "It seems to have been reserved to the people of this country," Alexander Hamilton wrote in the third sentence of *The Federalist Papers,* "by their conduct and their example, to decide the important question, whether societies of men are really capable or not, of establishing good government from reflection and choice."[10] The design of the Constitution that Madison, Hamilton, and Jay advocated in *The Federalist Papers* aims to achieve an equilibrium that would, as Gay puts it, "guard the passions of individuals for the sake of order and guard the guardians for the sake of freedom."[11] Preeminent among the European *philosophes* who influenced them was Montesquieu. Experience had shown that every man invested with power is apt to abuse it, he wrote in *De l'esprit des lois*. To prevent this abuse, Montesquieu argued that power should check power, a principle embodied by the Framers in the architecture of the American Constitution. What happens when those checks and the separation of powers are circumvented, corrupted, or in some other way nullified? The framers recognized that the mechanics and disposition of powers arranged by the Constitution are not its ultimate guardian. "If it be true that all governments rest on opinion," Madison wrote in *Federalist* No. 49, then the Constitution's ultimate protection resides in the First Amendment's prohibition against any law abridging freedom of speech.[12]

Freedom of speech can be abridged in ways the First Amendment does not protect. Speech can be rendered mute by the spiral of silence. Constitutional safeguards can be overcome by climate change's appeal to the great category of blackmail. Avoiding a planetary catastrophe requires action. Even if Congress expressly declines to legislate, global warming compels the executive to regulate. But it demands more. It demands that citizens agree—and requires of those who don't to hold their tongue, for it craves conformity and the spurious legitimacy of the approbation of the citizenry mass. It therefore requires dissent be silenced.

Thus, global warming harbors a strong impulse toward the governing modes of the absolutist and the political culture of the totalitarian. It is why global warming threatens the American republic, not on fabricated environmental and public health grounds, but to the roots of its being in an attack on the ideals conceived at its creation. The alignment of the natural bias of the media with the minatory power of the state means dissidents can be threatened, belittled, and demonized without the need to repeal the First Amendment—only for as long as the spiral of silence continues to do its work. Its triumph is not inevitable. Like their forefathers who won their freedom from an empire, Americans have it within their power to recover the ideals on which their republic is founded. They can have the courage to break the spiral of silence and restore the foundations of their republic to its timeless ideal.

Notes

1. America in Lilliput

1 Jeffrey Goldberg, "The Obama Doctrine," *The Atlantic*, April 2016, http://www.theatlantic.com/magazine/archive/2016/04/the-obama-doctrine/471525/ (accessed March 14, 2016).
2 Roland Huntford, *The New Totalitarians* (London, 1971), p. 20.
3 Ibid., p. 37.
4 Carmen M. Reinhart and Kenneth S. Rogoff, *This Time Is Different* (Princeton and Oxford, 2009), p. xxxvii.

2. The Great Transformation

1 UN Regional Information Centre for Western Europe, "Figueres: First Time the World Economy Is Transformed Intentionally," February 3, 2015, http://www.unric.org/en/latest-un-buzz/29623-figueres-first-time-the-world-economy-is-transformed-intentionally.
2 BBC News Magazine, "How Parasites Manipulate Us," February 19, 2014, http://www.bbc.co.uk/news/magazine-26240297.
3 Barack Obama, State of the Union address, January 20, 2015, http://www.whitehouse.gov/thepressoffice/2015/01/20/remarks-president-state-union-address-january-20-2015.
4 John C. Fyfe, Nathan P. Gillett, and Francis W. Zwiers, "Overestimated Global Warming over the Past 20 Years," *Nature Climate Change*, Vol. 3 (September 2013), Fig. 2b.
5 Peter W. Huber and Mark P. Mills, *The Bottomless Well: The Twilight of Fuel, the Virtue of Waste, and Why We Will Never Run out of Energy* (New York, 2006), p. 58.
6 John F. Wasik, *The Merchant of Power: Sam Insull, Thomas Edison, and the Creation of the Modern Metropolis* (New York, 2006), p. 19.
7 Thomas P. Hughes, *Networks of Power: Electrification in Western Society 1880–1930* (Baltimore, 1983), p. 22.
8 Tom McNichol, *AC/DC: The Savage Tale of the First Standards War* (San Francisco, 2006), p. 66.
9 John F. Wasik, *The Merchant of Power: Sam Insull, Thomas Edison, and the Creation of the Modern Metropolis* (New York, 2006), p. 122.
10 WBGU, *Serving Global Change Politics* (2013), p. 3.

11 Edgar Gärtner email to author, March 2, 2015.
12 As at December 31, 2009.
13 Accessed via http://notrickszone.com/2013/09/08/new-film-shows-hans-schellnhuber-claiming-himalayan-2035-glacier-melt-was-very-easy-to-calculate/#sthash.byCjxIqf.dpbs.
14 Pierre Gosselin, "Teutonic Power Grab . . . Schellnhuber and Co. Tell World to Do as They Say, or Globe Gets 230-Foot Sea Level Rise!" posted on NoTricksZone, December 8, 2014, http://notrickszone.com/2014/12/08/teutonic-power-grab-schellnhuber-co-tell-world-to-do-as-they-say-or-globe-gets-230-foot-sea-level-rise/#sthash.9sRTbCSM.dpbs.
15 Klimaretter.info, "'Klimapapst' im Umweltausschuss" ("Climate Pope" at the Environment Committee), December 3, 2014, http://www.klimaretter.info/politik/nachricht/17724-qklimapapstq-im-umweltausschuss.
16 Cora Stephan, "Ökodiktatoren" (Eco-dictators) posted on Die Achse des Guten, June 6, 2011, http://www.achgut.com/dadgdx/index.php/dadgd/article/oekodi/.
17 Potsdam Institute for Climate Impact Research, "Hans Joachim Schellnhuber Awarded Wilhelm Foerster Prize 2013," March 28, 2013, https://www.pik-potsdam.de/news/in-short/wilhelm-foerster-preis-2013-fuer-hans-joachim-schellnhuber?set_language=en.
18 Thomas Homer-Dixon, "The Great Transformation: Climate Change as Cultural Change," June 8, 2009, http://www.homerdixon.com/2009/06/08/the-great-transformation-climate-change-as-cultural-change/.
19 Edgar Gärtner contemporaneous notes.
20 Andreas Ernst, "Action Failure, Action Opportunities, and Action Success—A Psychological Perspective," June 10, 2009.
21 Edgar Gärtner contemporaneous notes.
22 Jeevan Vasagar, "Academic Linked to Gaddafi's Fugitive Son Leaves LSE," guardian.com, October 31, 2011, http://www.theguardian.com/education/2011/oct/31/saif-gaddafi-lse-academic.
23 David Held, "Democracy, Climate Change and Global Governance" (June 2009).
24 William J. Antholis, "The Good, the Bad, and the Ugly: EU-U.S. Cooperation on Climate Change," June 10, 2009, http://www.brookings.edu/research/speeches/2009/06/10-climate-antholis.
25 Author interview with Edgar Gärtner, February 24, 2015.
26 John D. Podesta, "The Great Transformation: Climate Change as Cultural Change," June 9, 2009, https://cdn.americanprogress.org/wp-content/uploads/issues/2009/06/pdf/podesta_germany1.pdf.
27 U.S. Energy Information Administration, *Electric Power Annual 2012* (December 2013), Table 4.3.
28 Edgar Gärtner contemporaneous notes.
29 Bernhard Pötter, "Klimapolitik verteilt das Weltvermögen neu" (Climate policy to redistribute global wealth), *Neue Zürcher Zeitung*, November 14, 2010.

30 Author interview with Edgar Gärtner, February 24, 2015.
31 Office of the Press Secretary, the White House, "Fact Sheet: US-China Joint Announcement on Climate Change and Clean Energy Co-operation," November 11, 2014, http://www.whitehouse.gov/the-press-office/2014/11/11/fact-sheet-us-china-joint-announcement-climate-change-and-clean-energy-c.
32 Author interview with Edgar Gärtner, February 24, 2015.
33 Jürgen Polzin and Christopher Onkelbach, "Notbremse" (Emergency brake), *Westdeutsche Allgemeine Zeitung*, June 10, 2009.
34 Moritz Hartmann, "Demnächst Biokirschen aus Timbuktu?" (Organic Cherries Coming Soon from Timbuktu?) *Frankfurter Allgemeine Zeitung*, June 17, 2009.
35 Ulli Kulke, "Klimaschutz killt die Demokratie," *Die Welt*, December 14, 2009.
36 Claus Leggewie, "Warnungen vor einer Ökodiktatur? Lächerlich!" (Eco-dictatorship ahead? Ridiculous!), *Die Welt*, May 25, 2011.
37 Edgar Gärtner, "Ökodiktatur auf Samtpfoten?" (Eco-dictatorship by stealth?) gaertner-online.de, May 27, 2011, http://gaertner-online.de/2011/05/27/okodiktatur-auf-samtpfoten/.
38 WBGU, *World in Transition: A Social Contract for Sustainability: Summary for Policy-Makers* (2011), p. 5.
39 Ibid., p. 1.
40 Ibid., p. 4.
41 Ibid., p. 4.
42 Ibid., p. 8.
43 Ibid., p. 6.
44 Ibid., p. 2.

3. Northern Lights

1 Roland Huntford, *The New Totalitarians* (London, 1971), p. 66.
2 Gustaf Arrhenius, Karin Caldwell, and Svante Wold, *A Tribute to the Memory of Svante Arrhenius (1859 –1927): A scientist Ahead of his Time* (Stockholm, 2008), p. 36.
3 Maria Björkman and Sven Widmalm "Selling Eugenics: The Case of Sweden," *Notes and Records of the Royal Society*, August 18, 2010.
4 Paul O'Mahony, "Sweden's 'Dark Legacy' Draws Crowds to Museum," The Local.se, January 9, 2007, http://www.thelocal.se/20070109/6041.
5 Roland Huntford, *The New Totalitarians* (London, 1971), pp. 9–10.
6 Ibid., p. 10.
7 Gunnar Broberg and Mattias Tydén, "Eugenics in Sweden: Efficient Care" in Gunnar Broberg and Nils Roll-Hansen, *Eugenics and the Welfare State: Sterilization Policy in Denmark, Sweden, Norway, and Finland* (East Lansing, 2005), p. 104.
8 Stellan Andersson, "On the Value of Personal Archives: Some Examples from the Archives of Alva and Gunnar Myrdal—with a Main Focus on

Gunnar," http://edoc.hu-berlin.de/nordeuropaforum/1999-1/andersson-stellan-15/XML/ (accessed March 3, 2015).

9 Gunnar Broberg and Mattias Tydén, "Eugenics in Sweden: Efficient Care" in Gunnar Broberg and Nils Roll-Hansen, *Eugenics and the Welfare State: Sterilization Policy in Denmark, Sweden, Norway, and Finland* (East Lansing, 2005), p. 97.

10 Örjan Appelqvist, *The Political Economy of Gunnar Myrdal: Transcending Dilemmas Post-2008* (London and New York, 2013), p. 1.

11 Gunnar Myrdal, "Prize Lecture: The Equality Issue in World Development," March 17, 1975, http://www.nobelprize.org/nobel_prizes/economic-sciences/laureates/1974/myrdal-lecture.html.

12 Jay Nordlinger, *Peace, They Say; A History of the Nobel Peace Prize, the Most Famous and Controversial Prize in the World* (New York and London, 2012), p. 245.

13 Gunnar Broberg and Mattias Tydén, "Eugenics in Sweden: Efficient Care" in Gunnar Broberg and Nils Roll-Hansen, *Eugenics and the Welfare State: Sterilization Policy in Denmark, Sweden, Norway, and Finland* (East Lansing, 2005), p. 105.

14 Roland Huntford, *The New Totalitarians* (London, 1971), pp. 62–63.

15 Gunnar Broberg and Mattias Tydén, "Eugenics in Sweden: Efficient Care" in Gunnar Broberg and Nils Roll-Hansen, *Eugenics and the Welfare State: Sterilization Policy in Denmark, Sweden, Norway, and Finland* (East Lansing, 2005), pp. 114–115.

16 Ibid., p. 115.

17 Ibid., p. 107.

18 Ibid., p. 108.

19 Gunnar Broberg and Nils Roll-Hansen, *Eugenics and the Welfare State: Sterilization Policy in Denmark, Sweden, Norway, and Finland* (East Lansing, 2005), pp. ix and xi.

20 Ibid., p. 122.

21 Joachim Israel, "An Excerpt from *Ett upproriskt liv* (A rebellious life) in Peter Stenberg (Ed.) *Contemporary Jewish Writing in Sweden: An Anthology* (Lincoln, 2004), p. 215.

22 Ibid., p. 215.

4. Europe's First Greens

1 Adolf Hitler, *Mein Kampf,* http://www.bibliotecapleyades.net/archivos_pdf/meinkampf.pdf (accessed, March 23, 2015), p. 147.

2 Adam Tooze, *The Wages of Destruction: The Making and Breaking of the Nazi Economy* (New York, 2006), p. xxiv.

3 Robert N. Proctor, *The Nazi War on Cancer* (Princeton, 1999), p. 7.

4 Janet Biehl and Peter Staudenmaier, *Ecofascism: Lessons from the German Experience* (Edinburgh, 1995), p. 1.

5 Adolf Hitler, *Mein Kampf,* http://www.bibliotecapleyades.net/archivos_pdf/meinkampf.pdf (accessed March 23, 2015), p. 147.
6 Ronald L. Meek, *Marx and Engels on Malthus* (London, 1953), p. 63.
7 Peter Staudenmaier, "Fascist Ecology: The "Green Wing" of the Nazi Party and Its Historical Antecedents" in Janet Biehl and Peter Staudenmaier, *Ecofascism: Lessons from the German Experience* (Edinburgh, 1995), p. 4.
8 Robert N. Proctor, *The Nazi War on Cancer* (Princeton, 1999), p. 15.
9 Richard Weikart, *Hitler's Ethic: The Nazi Pursuit of Evolutionary Progress* (New York, 2009), p. 14.
10 Warren T. Reich, "The Care-Based Ethic of Nazi Medicine and the Moral Importance of What We Care About," *The American Journal of Bioethics,* Vol. 1, No. 1, (Winter 2001), pp. 65–66; Robert N. Proctor, *The Nazi War on Cancer* (Princeton, 1999), p. 24.
11 Ludwig Klages (tr. Joe Pryce), *Man and Earth* (1913), http://www.revilooliver.com/Writers/Klages/Man_and_Earth.html (accessed 24th March, 2015).
12 Ibid.
13 Peter Staudenmaier, "Fascist Ecology: The 'Green Wing' of the Nazi Party and Its Historical Antecedents" in Janet Biehl and Peter Staudenmaier, *Ecofascism: Lessons from the German Experience* (Edinburgh, 1995), p. 12.
14 Adolf Hitler, *Mein Kampf,* http://www.bibliotecapleyades.net/archivos_pdf/meinkampf.pdf (accessed March 23, 2015), p. 111.
15 Ibid., p. 109.
16 Adam Tooze, *The Wages of Destruction: The Making and Breaking of the Nazi Economy* (New York, 2006), p. xxiv.
17 Adolf Hitler, *Mein Kampf,* http://www.bibliotecapleyades.net/archivos_pdf/meinkampf.pdf (accessed March 23, 2015), p. 109.
18 *Völkischer Beobachter,* "Der Weg zur nationalen Kraftwirtschaft," February 24, 1932.
19 https://docs.google.com/viewer?url=patentimages.storage.googleapis.com/pdfs/US1213889.pdf.
20 Franz Lawaczeck, *Technik und Wirtschaft im Dritten Reich: Ein Arbeitsbeschaffungsprogramm* (3rd edition, Nationalsozialistische Bibliotek, 1933), p. 7.
21 Ibid., p. 17.
22 Ibid., p. 37.
23 Ibid., pp. 47–48.
24 Ibid., pp. 60–61.
25 Ibid., p. 54.
26 Ibid., p. 12.
27 Royal Academy of Engineering, *Wind Energy: Implications of Large-Scale Deployment on the GB Electricity System* (London, 2014), p. 9.

28 Michael J. Kelly, *Technology Introduction in the Context of Decarbonization: Lessons from Recent History* (GWPF Note 7), Table 1.
29 Franz Lawaczeck, *Technik und Wirtschaft im Dritten Reich: Ein Arbeitsbeschaffungsprogramm* (3rd edition, Munich, 1933), p. 48.
30 Matthias Heymann, *Die Geschichte der Windenergienutzung 1890–1990* (Frankfurt, 1995), p. 182.
31 Ibid., p. 186.
32 Ibid., p. 193.
33 Hermann Rauschning, *Hitler Speaks* (Third Impression, December 1939, Andover), p. 33.
34 Ibid., p. 34.
35 Ibid., p. 34.
36 H. R. Trevor-Roper and Gerhard L. Weinberg, *Hitler's Table Talk 1941–1944: Secret Conversations* (New York, 2000), p. 20.
37 Matthias Heymann, *Die Geschichte der Windenergienutzung 1890–1990* (Frankfurt, 1995), pp. 193–194.
38 Ibid., p. 198.
39 See http://hermann-foettinger.de/Projekte/RAW/raw_schriften.pdf (accessed April 9, 2015).
40 Robert N. Proctor, *The Nazi War on Cancer* (Princeton, 1999), pp. 131–132.
41 Ibid., p. 138.
42 Rupert Darwall, *The Age of Global Warming: A History* (London, 2013), pp. 40–41.
43 Robert N. Proctor, *The Nazi War on Cancer* (Princeton, 1999), p. 120 and p. 124.
44 Ibid., p. 74.
45 Ibid., pp. 134–135.
46 Ibid., p. 219.
47 Ibid., Table 6.1 and p. 270.
48 Adam Tooze, *The Wages of Destruction: The Making and Breaking of the Nazi Economy* (New York, 2006), p. 116.
49 Thomas Hager, *The Alchemist of Air: A Jewish Genius, a Doomed Tycoon, and the Scientific Discovery that Fed the World but Fueled the Rise of Hitler* (New York, 2009), p. 226.
50 Ibid., p. 231.
51 Holocaust Education and Archive Research Team, "I.G. Farbenindustrie AG German Industry and the Holocaust," http://www.holocaustresearchproject.org/economics/igfarben.html (accessed March 27, 2015).
52 Adam Tooze, *The Wages of Destruction: The Making and Breaking of the Nazi Economy* (New York, 2006), p. 117.
53 Erhard Schütz and Eckhard Gruber, *Mythos Reichsautobahn: Bau und Inszenierung der "Strassen des Führers" 1933–1941* (Berlin, 1996), p. 142.

54 Adam Tooze, *The Wages of Destruction: The Making and Breaking of the Nazi Economy* (New York, 2006), p. 150.
55 Ibid., p. 154.
56 Ibid., p. 150.
57 Ibid., p. 46.
58 Peter Staudenmaier, "Fascist Ecology: The "Green Wing" of the Nazi Party and Its Historical Antecedents" in Janet Biehl and Peter Staudenmaier, *Ecofascism: Lessons from the German Experience* (Edinburgh, 1995), pp. 20–21.
59 Rupert Darwall, *The Age of Global Warming: A History* (London, 2013), p. 40.
60 Adam Tooze, *The Wages of Destruction: The Making and Breaking of the Nazi Economy* (New York, 2006), p. 652.
61 Robert N. Proctor, *The Nazi War on Cancer* (Princeton, 1999), p. 15.
62 Rachel Carson, *Silent Spring* (London, 2000), p. 213.
63 Ibid., p. 212.
64 Ibid., p. 25.
65 Ibid., p. 168.
66 Richard Doll and Richard Peto, "The Quantitative of Cancer Causes of Cancer: Estimates of Avoidable Risks in the United States Today," *Journal of the National Cancer Institute,* Vol. 66, No. 6 (June 1981), pp. 1194–1308.
67 Robert N. Proctor, *The Nazi War on Cancer* (Princeton, 1999), pp. 248–249 and 277.
68 Ibid., p. 249.
69 Peter Staudenmaier, "Fascist Ecology: The "Green Wing" of the Nazi Party and Its Historical Antecedents" in Janet Biehl and Peter Staudenmaier, *Ecofascism: Lessons from the German Experience* (Edinburgh, 1995), p. 24.
70 Thomas Rohrkrämer, "Antimodernism, Reactionary Modernism and National Socialism," *Contemporary European History,* Vol. 8, No. 1 (1999), p. 50.

5. Intellectuals, Activists, and Experts

1 Joseph A. Schumpeter, *Capitalism, Socialism and Democracy* (First published 1943, 1981 edition, London), p. 130.
2 Jean-Louis Ferrier, Jacques Boetsch, and Françoise Giroud (tr. Helen Weaver), "Marcuse Defines His New Left Line," *New York Times,* October 27, 1968.
3 Saul Alinsky, *Rules For Radicals: A Practical Primer for Realistic Radicals,* (New York, 1971), p. 128.
4 Gordon Fraser, "How Nazi Germany Lost Its Nuclear Edge," thejc.com, October 25, 2012, http://www.thejc.com/comment-and-debate/comment/88125/how-nazi-germany-lost-its-nuclear-edge (accessed May 1, 2015).
5 Ibid.

6 Richard Swedberg, *Joseph A. Schumpeter: His Life and Work* (Cambridge, 1991), p. 150.
7 Joseph A. Schumpeter, *Capitalism, Socialism and Democracy* (first published 1943, 1981 edition, London), p. 83.
8 Ibid., p. 134.
9 Ibid., p. 143.
10 Ibid., p. 146.
11 Ibid., p. 151.
12 Richard Swedberg, *Joseph A. Schumpeter: His Life and Work* (Cambridge, 1991), p. 150.
13 Martin Jay, *The Dialectical Imagination: A History of the Frankfurt School and the Institute of Social Research 1932–1950* (London, 1973), p. 8.
14 Ibid., p. 46.
15 Ibid., p. 63.
16 Jean-Louis Ferrier, Jacques Boetsch, and Françoise Giroud (tr. Helen Weaver), "Marcuse Defines His New Left Line," *New York Times*, October 27, 1968.
17 Herbert Marcuse, "Ecology and the Critique of Modern Society," *Capital Nature Socialism*, March 1992, p. 33.
18 Martin Jay, *The Dialectical Imagination: A History of the Frankfurt School and the Institute of Social Research 1932–1950* (London, 1973), p. 63.
19 Ibid., p. 20.
20 Kenneth A. Briggs, "Marcuse, Radical Philosopher, Dies," *New York Times*, July 31, 1979.
21 Jean-Louis Ferrier, Jacques Boetsch, and Françoise Giroud (tr. Helen Weaver), "Marcuse Defines His New Left Line," *New York Times*, October 27, 1968.
22 Martin Jay, *The Dialectical Imagination: A History of the Frankfurt School and the Institute of Social Research 1932–1950* (London, 1973), pp. 32–33.
23 Thomas Wheatland, *The Frankfurt School in Exile* (Minneapolis, 2009), p. 255.
24 Ibid., p. 256.
25 Edward Shils, "Daydreams and Nightmares: Reflections on the Criticism of Mass Culture," *Sewanee Review*, Vol. 65, No. 4 (Autumn 1957), p. 596.
26 Martin Jay, *The Dialectical Imagination: A History of the Frankfurt School and the Institute of Social Research 1932–1950* (London, 1973), p. 57.
27 Edward Shils, "Daydreams and Nightmares: Reflections on the Criticism of Mass Culture," *Sewanee Review*, Vol. 65, No. 4 (Autumn 1957), p. 600.
28 Ibid., p. 590.
29 Paul Hockenos, *Joshka Fischer and the Making of the Berlin Republic: An Alternative History of Postwar Germany* (Oxford, 2008), p. 60.
30 Ibid.
31 Stephen R. C. Hicks, *Explaining Postmodernism: Skepticism and Socialism from Rousseau to Foucault* (Phoenix and New Berlin, WI, 2004), p. 159.

32 Herbert Marcuse, "Repressive Tolerance," in Robert P. Wolff, Barrington Moore Jr., and Herbert Marcuse, *A Critique of Pure Tolerance* (London, 1969), p. 98.
33 Ibid., p. 125.
34 Ibid., p. 119.
35 Herbert Marcuse, "Repressive Tolerance," in Robert P. Wolff, Barrington Moore Jr., and Herbert Marcuse, *A Critique of Pure Tolerance* (London, 1969), p. 95.
36 Jean-Louis Ferrier, Jacques Boetsch, and Françoise Giroud (tr. Helen Weaver), "Marcuse Defines His New Left Line," *New York Times*, October 27, 1968.
37 Thomas Wheatland, *The Frankfurt School in Exile* (Minneapolis, 2009), pp. 340–341.
38 Ibid., p. 287.
39 Rudi Dutschke, "On Anti-Authoritarianism," in Carl Oglesby (ed.), *The New Left Reader* (Boston, 1968), p. 249.
40 Jean-Louis Ferrier, Jacques Boetsch, and Françoise Giroud (tr. Helen Weaver), "Marcuse Defines His New Left Line," *New York Times*, October 27, 1968.
41 Herbert Marcuse, "Ecology and the Critique of Modern Society," *Capital Nature Socialism*, March 1992, p. 33.
42 Elinor Ostrom, Joanna Burger, Christopher B. Field, Richard B. Norgaard, and David Policansky, "Revisiting the Commons: Local Lessons, Global Challenges," *Science*, Vol. 284, No. 5412 (April 9, 1999), pp. 278–282.
43 Garrett Hardin, "The Tragedy of the Commons," *Science*, Vol. 162, No. 3859 (December 13, 1968), pp. 1243–1248.
44 Garrett Hardin, *Biology: Its Principles and Implications* (2nd ed., San Francisco, 1966), p. 706.
45 Ibid., p. 707.
46 Ibid., p. 706.
47 Garrett Hardin, *Living within Limits: Ecology, Economics and Population Taboos* (New York and Oxford, 1993), p. 270.
48 Garrett Hardin, *Biology: Its Principles and Implications* (2nd ed., San Francisco, 1966), p. 6.
49 Ibid., p. 123.
50 Ibid., p. 294.
51 Ibid., p. 131.
52 See, for example, Anna Bramwell, *Blood and Soil: Richard Walter Darré and Hitler's "Green Party"* (Bourne End, 1985), pp. 49–50.
53 Garrett Hardin, *Biology: Its Principles and Implications* (2nd ed., San Francisco, 1966), p. 6.
54 Murray Bookchin, "Ecology and Revolutionary Thought" (1964), http://dwardmac.pitzer.edu/Anarchist_Archives/bookchin/ecologyandrev.html (accessed July 15, 2015).

55 C. Wright Mills, "Letter to the New Left," *New Left Review*, No. 5, September–October 1960, https://www.marxists.org/subject/humanism/mills-c-wright/letter-new-left.htm (accessed May 13, 2015).
56 Ibid.
57 C. Wright Mills, *The Power Elite* (Oxford and New York, 1999), p. 4.
58 Students for a Democratic Society, "Port Huron Statement of the Students for a Democratic Society" (1962), http://coursesa.matrix.msu.edu/~hst306/documents/huron.html (accessed May 13, 2015).
59 Joseph A. Schumpeter, *Capitalism, Socialism and Democracy* (First published 1943, 1981 edition, London), p. 143.
60 Jarol Manheim, in *Biz-War and the Out-of-Power Elite: The Progressive-Left Attack on the Corporation* (Mahwah, NJ, 2004), p. 84.
61 Sidney Hyman, *The Aspen Idea* (Norman, OK, 1975), p. 159.
62 Donald Gibson, *Environmentalism and Power* (Huntington, NY, 2002), p. 64.
63 Sidney Hyman, *The Aspen Idea* (Norman, OK, 1975), p. 268.
64 James Reston, "Apsen, Colo.: The Philosophers at Bay," *New York Times*, September 2, 1970.
65 Peter Passell, Marc Roberts and Leonard Ross, "The Limits to Growth; A Report for the Club of Rome's Project on the Predicament of Mankind." *New York Times*, April 2, 1972.

6. Raindrops

1 Naomi Oreskes and Erik Conway, *Merchants of Doubt: How a Handful of Scientists Obscured the Truth on Issues from Tobacco Smoke to Global Warming* (New York, Berlin, London, 2010), p. 69.
2 Michael E. Kowalok, "Commons Threads," *Environment*, Vol. 35, No. 6 (July/August 1993).
3 Barbara Ward and René Dubos, *Only One Earth: The Care and Maintenance of a Small Planet* (first published, 1972, London, 1974), pp. 100 and 265–270.
4 Naomi Oreskes and Erik Conway, *Merchants of Doubt: How a Handful of Scientists Obscured the Truth on Issues from Tobacco Smoke to Global Warming* (New York, Berlin, London, 2010), p. 76.
5 Edward C. Krug, "The Great Acid Rain Flimflam," in Jay H. Lehr (ed.), *Rational Readings on Environmental Concerns* (New York, 1992), p. 35.
6 Royal Ministry of Foreign Affairs, Royal Ministry of Agriculture, *Air Pollution across National Boundaries—The Impact on the Environment of Sulphur in Air and Precipitation: Sweden's Case Study for the United Nations Conference on the Human Environment* (Stockholm, 1971), p. 7.
7 Ibid., p. 57.
8 Ibid., p. 59.
9 Aaron Wildavsky, *But Is It True?: A Citizen's Guide to Environmental Health and Safety Issues* (Cambridge, MA, and London, 1995), p. 296.

10 Royal Ministry of Foreign Affairs, Royal Ministry of Agriculture, *Air Pollution across National Boundaries—The Impact on the Environment of Sulphur in Air and Precipitation: Sweden's Case Study for the United Nations Conference on the Human Environment* (Stockholm, 1971), p. 9.
11 Ibid., p. 57.
12 Ibid., p. 85.
13 Ibid.
14 Ibid., p. 9.
15 Olof Palme, "Statement by Prime Minister Olof Palme in the Plenary Meeting, June 6, 1972, pp. 8–9.
16 Bo J. Theutenberg interview with author, June 8, 2015.
17 Bill Mayr, "Remembering Olof Palme," *Kenyon College Alumni Bulletin*, Vol. 34, No. 2 (Winter 2012), http://bulletin.kenyon.edu/x3901.xml (accessed April 14, 2015).
18 Bo J. Theutenberg, *Dagbok Från UD* Vols. 1 and 2 (Skara, 2012, 2014).
19 Ibid., pp. 68 and 76 *et seq.*
20 Roland Huntford, *The New Totalitarians* (London, 1971), p. 204.
21 Ibid., p. 215.
22 Ibid., p. 288.
23 Bo J. Theutenberg, *Dagbok Från UD 1962–1976*, Vol. 2 (Skara, 2014), pp. 248–253.
24 Bo J. Theutenberg interview with author, June 8, 2015.
25 UNEP, "Declaration of the United Nations Conference on the Human Environment," June 16, 1972, http://www.unep.org/Documents.Multilingual/Default.asp?documentid=97&articleid=1503 (accessed May 27, 2015).
26 Sonja Boehmer-Christiansen and Jim Skea, *Acid Politics: Environmental and Energy Policies in Britain and Germany* (London and New York, 1991), p. 27.
27 UNECE, 1979 Convention on Long-Range Transboundary Air Pollution, Article 2.
28 Sonja Boehmer-Christiansen and Jim Skea, *Acid Politics: Environmental and Energy Policies in Britain and Germany* (London and New York, 1991), p. 27.
29 Ibid., p. 190.
30 Ibid., p. 28.
31 UNECE, Protocol to the 1979 Convention on Long-Range Transboundary Air Pollution on the Reduction of Sulphur Emissions or Their Transboundary Fluxes by at Least 30 Per Cent, Article 2.
32 Nils Roll-Hansen, *Ideological Obstacles to Scientific Advice in Politics? The Case of "Forest Rain" from "Acid Rain"* (November 2002), p. 4.
33 Ibid., p. 7.
34 Ibid., p. 9.
35 National Environmental Protection Board, *Acid Magazine*, No. 8 (September 1989), p. 5.

36 Perry Johanson, *Saluting the Yellow Emperor: A Case of Swedish Sinography* (Leiden, 2012), p. 163, fn 12.
37 Kerstin Österberg, "Skogin Dör I England" (The Wood Dies in England), *Ny Teknik*, 1985:39.
38 Margaret Thatcher, "Speech to Royal Society Dinner (marking end of Surface Water Acidification Programme)," March 22, 1990, http://www.margaretthatcher.org/document/108046 (accessed July 18, 2015).
39 Hans Lundberg author interview, June 9, 2015.
40 Jimmy Carter, "Environmental Priorities and Programs Message to the Congress," August 2, 1979, http://www.presidency.ucsb.edu/ws/?pid=32684 (accessed May 29, 2015).
41 Committee on the Atmosphere and the Biosphere, Board on Agriculture and Renewable Resources, Commission on Natural Resources, National Research Council, *Atmosphere-Biosphere Interactions: Toward a Better Understanding of the Ecological Consequences of Fossil Fuel Combustion* (Washington, D.C., 1981), p. 1.
42 Ibid., pp. 3–4.
43 Ibid., p. 2.
44 Ibid.
45 Ibid., p. 6.
46 Ibid., p. 3.
47 Ibid., p. 7.
48 Ibid., p. 6.

7. Acid Denial

1 Aaron Wildavsky, *But Is It True?: A Citizen's Guide to Environmental Health and Safety Issues* (Cambridge, MA, and London, 1995), p. 274.
2 Robert B. Stewart, "Negotiations on Acid Rain," in Jurgen Schmandt, Judith Clarkson, and Hillard Roderick (eds.), *Acid Rain and Friendly Neighbors: The Policy Dispute between Canada and the United States* (Revised edition, Durham, 1988), p. 74.
3 Robert B. Stewart, "Negotiations on Acid Rain," in Jurgen Schmandt, Judith Clarkson, and Hillard Roderick (eds.), *Acid Rain and Friendly Neighbors: The Policy Dispute between Canada and the United States* (Revised edition, Durham, 1988), p. 69.
4 Ibid., p. 74.
5 William Nierenberg et al., *Report of the Acid Rain Peer Review Panel* (July 1984), p. iii–6.
6 Ibid., p. v.
7 Ibid., pp. ii–iii.
8 Ibid., p. ii–2.
9 Ibid., pp. iv–v.
10 S. Fred Singer email to author, September 27, 2016.

11 William Nierenberg et al., *Report of the Acid Rain Peer Review Panel* (July 1984), p.A5–10
12 Ibid., p. A5–8, fn.3
13 Ibid., p. A5–7
14 Ibid., pp. A4–3, A4–5
15 Edward C. Krug and Charles R. Fink, "Acid Rain on Acid Soil: A New Perspective," *Science*, New Series, Vol. 221, No. 4610 (August 5, 1983), p. 520.
16 Ibid., p. 524.
17 Ibid.
18 Edward C. Krug, "The Great Acid Rain Flimflam," in Jay H. Lehr (ed.), *Rational Readings on Environmental Concerns* (New York, 1992), p. 36.
19 Kim J. DeRidder, "The Nature and Effects of Acid Rain: A Comparison of Assessments," in Jurgen Schmandt, Judith Clarkson, and Hillard Roderick (eds.), *Acid Rain and Friendly Neighbors: The Policy Dispute between Canada and the United States* (Revised edition, Durham, 1988), p. 62.
20 Philip Shabecoff, "Government Acid Rain Report Comes under Sharp Attack," *New York Times*, September 22, 1987.
21 Philip Shabecoff, "Study Discounts Immediate Peril of Acid Rain," *New York Times*, September 18, 1987.
22 President George H. W. Bush, "Remarks Announcing Proposed Legislation to Amend the Clean Air Act," June 12, 1989, http://www.presidency.ucsb.edu/ws/?pid=17134 (accessed June 1, 2015).
23 Patricia M. Irving (ed.), *Acidic Deposition: State of Science and Technology: Summary Report of the US National Acid Precipitation Assessment Program* (Washington, D.C., September 1991), p. 135.
24 Ibid., p. 144.
25 Ibid.
26 Ibid., p. 129.
27 Edward C. Krug, "The Great Acid Rain Flimflam," in Jay H. Lehr (ed.), *Rational Readings on Environmental Concerns* (New York, 1992), p. 38.
28 Patricia M. Irving (ed.), *Acidic Deposition: State of Science and Technology: Summary Report of the US National Acid Precipitation Assessment Program* (Washington, D.C., September 1991), p. 208.
29 Sonja Boehmer-Christiansen and Jim Skea, *Acid Politics: Environmental and Energy Policies in Britain and Germany* (London and New York, 1991), p. 199.
30 Ibid.
31 European Commission, "Large Combustion Plants Directive," http://ec.europa.eu/environment/industry/stationary/lcp/legislation.htm (accessed June 2, 2015).
32 Aaron Wildavsky, *But Is It True?: A Citizen's Guide to Environmental Health and Safety Issues* (Cambridge, MA, and London, 1995), p. 302.

33 Howard Kurtz, "Is Acid Rain a Tempest In News Media Teapot?" *Washington Post*, January 14, 1991.
34 Robert W. McGee, "Some Thoughts on Acid Rain, the Clean Air Act and Implications for Acid Rain Policy in the USA and Europe," July 1996, http://www.researchgate.net/profile/Robert_Mcgee2/publication/228282122_Some_Thoughts_on_Acid_Rain_the_Clean_Air_Act_and_Implications_for_Acid_Rain_Policy_in_the_USA_and_Europe/links/00b4952039fba480fc000000.pdf (accessed June 2, 2015).
35 Howard Kurtz, "Is Acid Rain a Tempest In News Media Teapot?" *Washington Post*, January 14, 1991.
36 Warren Brookes, "Scientific McCarthyism at the EPA?," *Washington Times*, May 1, 1991.
37 Howard Kurtz, "Is Acid Rain a Tempest In News Media Teapot?," *Washington Post*, January 14, 1991.
38 William Anderson, "Acid Test," *Reason*, January 1992, http://sppiblog.org/tag/william-l-anderson (accessed June 2, 2015).
39 Ibid.
40 Warren Brookes, "Scientific McCarthyism at the EPA?," *Washington Times*, May 1, 1991.
41 William Anderson, "Acid Test," *Reason*, January 1992, http://sppiblog.org/tag/william-l-anderson (accessed June 2, 2015).
42 Warren Brookes, "Scientific McCarthyism at the EPA?," *Washington Times*, May 1, 1991.
43 Dallas Burtraw, Alan Krupnick, Erin Mansur, David Austin, and Deirdre Farrell, "*The Costs and Benefits of Reducing Acid Rain* (Washington, D.C., September 1997), p. 8.
44 Ibid., Table 2.
45 L. C. Green and S. R. Armstrong, "Particulate Matter in Ambient Air and Mortality: Toxicologic Perspectives," *Regulatory Toxicology and Pharmacology*, Vol. 38, 326–335.
46 G. Koop, R. McKitrick, and Lise Tole, "Air Pollution, Economic Activity and Respiratory Illness: Evidence from Canadian Cities, 1974–1994," *Environmental Modelling and Software*, Vol. 25, No. 7 (July 2010), pp. 873–885.
47 Gene E. Likens, "The Science of Nature, the Nature of Science: Long-Term Ecological Studies at Hubbard Brook," *Proceedings of the American Philosophical Society*, Vol. 143, No. 4 (December 1999), p. 559.
48 IPCC, *Climate Change 2013 The Physical Science Basis: Summary for Policy Makers* (2013), p. 18.
49 EPA, http://www.epa.gov/acidrain/ (accessed June 3, 2015).
50 Karl Popper, *Conjectures and Refutations* (Abingdon and New York, 2002), p. 293.

8. Double Cross

1 Independent Commission on Disarmament and Security Issues under the Chairmanship of Olof Palme, *Common Security: A Programme for Disarmament* (London, 1982), p. vii.
2 Philip Short, *Mitterrand: A Study in Ambiguity* (London, 2013), p. 333.
3 Olof Palme, "Statement by Prime Minister Olof Palme in the Plenary Meeting," June 6, 1972, p. 6.
4 http://nome.unak.is/nm-marzo-2012/6-1x/24-articles61/77-olof-palme-one-life-many-readings (accessed April 14, 2015).
5 Franklin D. Scott, *Scandinavia* (Cambridge, MA, 1990), p. 292.
6 Roland Huntford, *The New Totalitarians* (London, 1971), p. 230.
7 Bill Mayr, "Remembering Olof Palme," *Kenyon College Alumni Bulletin* Vol. 34, No. 2 (Winter 2012), http://bulletin.kenyon.edu/x3901.xml (accessed April 14, 2015).
8 Roland Huntford, *The New Totalitarians* (London, 1971), p. 291.
9 Ibid., p. 341.
10 Ibid., p. 152.
11 Bo J. Theutenberg, *Dagbok Från UD 1962–1976*, Vol. 2 (Skara, 2014), p. 200.
12 Bo J. Theutenberg, *Dagbok Från UD 1981–1983*, Vol. 1 (Skara, 2012), pp. 291–299; and Bo J. Theutenberg, *Dagbok Från UD 1962–1976*, Vol. 2 (Skara, 2014), pp. 88, 98–99, and p. 198.
13 Bo J. Theutenberg, *Dagbok Från UD 1962–1976*, Vol. 2 (Skara, 2014), pp. 86–87.
14 Ibid., p. 197.
15 Ibid., pp. 86–87.
16 Barnaby Feder, "The Palme Style: A Strong Tilt to Foreign Policy," *New York Times*, March 4, 1986.
17 Örjan Berner, *Soviet Policies towards the Nordic Countries* (Center for International Affairs Harvard, May 1986), p. 133.

9. Born Again Greens

1 Werner Hülsberg, *The German Greens: A Social and Political Profile* (London and New York, 1988), p. 70.
2 Paul Berman, *Power and the Idealists* (New York and London, 2007), p. 73.
3 SPD, "Bad Godesberg Program" (November 1959), http://homepage2.nifty.com/socialist-consort/SDforeign/Godes/Godes.html#Preamble (accessed April 20, 2015).
4 Werner Hülsberg, *The German Greens: A Social and Political Profile* (London and New York, 1988), p. 25.
5 SPD, "Bad Godesberg Program" (November 1959), http://homepage2.nifty.com/socialist-consort/SDforeign/Godes/Godes.html#Preamble (accessed April 20, 2015).
6 Fritz Vahrenholt author interview, June 11, 2015.

7 Anna Bramwell, *The Fading of the Greens: The Decline of Environmental Politics in the West* (New Haven and London, 1994), p. 94.
8 Nicholas Kulish, "Spy Fired Shot That Changed West Germany," *New York Times*, May 27, 2009.
9 Paul Hockenos, *Joshka Fischer and the Making of the Berlin Republic: An Alternative History of Postwar Germany* (Oxford, 2008), p. 89.
10 Paul Berman, *Power and the Idealists* (New York and London, 2007), p. 54.
11 Paul Hockenos, *Joshka Fischer and the Making of the Berlin Republic: An Alternative History of Postwar Germany* (Oxford, 2008), p. 91.
12 Ibid., p. 96.
13 Ibid.
14 Ibid.
15 Christopher Andrew and Vasili Mitrokhin, *The Mitrokhin Archive: The KGB in Europe and the West* (London, 1999), p. 580.
16 Ibid., pp. 580–581.
17 Rudolf Bahro, *From Red to Green: Interviews with New Left Review* (translated by Gus Fagan and Richard Hurst, London, 1984), p. 42.
18 Ibid., p. 114.
19 Ibid., p. 105.
20 Paul Berman, *Power and the Idealists* (New York and London, 2007), p. 54.
21 Ibid., p. 21.
22 Paul Hockenos, *Joshka Fischer and the Making of the Berlin Republic: An Alternative History of Postwar Germany* (Oxford, 2008), p. 118.
23 Paul Berman, *Power and the Idealists* (New York and London, 2007), p. 49.
24 Paul Hockenos, *Joshka Fischer and the Making of the Berlin Republic: An Alternative History of Postwar Germany* (Oxford, 2008), p. 114.
25 Paul Berman, *Power and the Idealists* (New York and London, 2007), p. 21.
26 Paul Hockenos, *Joshka Fischer and the Making of the Berlin Republic: An Alternative History of Postwar Germany* (Oxford, 2008), p. 121.
27 Ibid., p. 124.
28 Paul Berman, *Power and the Idealists* (New York and London, 2007), p. 37.
29 Ibid., p. 61.
30 Werner Hülsberg, *The German Greens: A Social and Political Profile* (London and New York, 1988), p. 59.
31 Ibid., p. 69.
32 Author interview with Benny Peiser, April 16, 2015.
33 Paul Hockenos, *Joshka Fischer and the Making of the Berlin Republic: An Alternative History of Postwar Germany* (Oxford, 2008), p. 136.
34 Werner Hülsberg, *The German Greens: A Social and Political Profile* (London and New York, 1988), p. 60.
35 David S. Yost, "The Soviet Campaign against INF in West Germany," in Brian D. Dailey and Patrick J. Parker (ed.), *Soviet Strategic Deception* (Lexington and Toronto, 1987), p. 344.

36 Werner Hülsberg, *The German Greens: A Social and Political Profile* (London and New York, 1988), p. 103.
37 Benny Peiser interview with author, May 21, 2015.
38 John Vincour, "Antimissile Group in Bonn Is Divided," *New York Times*, April 6, 1982.
39 Werner Hülsberg, *The German Greens: A Social and Political Profile* (London and New York, 1988), p. 70.
40 Mark Hertgaard, "The Legacy of Petra K. Kelly" in Petra K. Kelly *Thinking Green: Essays on Environmentalism, Feminism, and Nonviolence* (Berkeley, 1994), p. 138.
41 James M. Markham, "For "Greens" It's Make Waves, Not War," *New York Times* October 3, 1982.
42 Timothy Garton Ash, *In Europe's Name: Germany and the Divided Continent* (London, 1993), p. 30.
43 *New York Times*, "Tenacious Leader of the Green Party—Petra Karin Kelly," *New York Times*, March 7, 1983.
44 Rudolf Bahro, *From Red to Green: Interviews with New Left Review* (translated by Gus Fagan and Richard Hurst, London, 1984), p. 192.
45 Isabel Hilton, "The Green with a smoking gun: There's no mystery about who killed Petra Kelly and her lover Gert Bastian, but nobody knows the reason. Isabel Hilton has investigated their deaths and wonders whether Green general had grey areas in his past," *The Independent*, April 21, 1994, http://www.independent.co.uk/life-style/the-green-with-a-smoking-gun-theres-no-mystery-about-who-killed-petra-kelly-and-her-lover-gert-bastian-but-nobody-knows-the-reason-isabel-hilton-has-investigated-their-deaths-and-wonders-whether-green-general-had-grey-areas-in-his-past-1372660.html (accessed April 21, 2015).
46 Vladimir Bukovsky, "The Peace Movement and the Soviet Union," *Commentary* May 1, 1982.
47 David S. Yost, "The Soviet Campaign against INF in West Germany," in Brian D. Dailey and Patrick J. Parker (ed.), *Soviet Strategic Deception* (Lexington and Toronto, 1987), p. 366.
48 Ibid., p. 346.

10. Scientists for Peace

1 Independent Commission on Disarmament and Security Issues under the Chairmanship of Olof Palme, *Common Security: A Programme for Disarmament* (London, 1982), p. 51.
2 Carl Sagan, "Nuclear War and Climatic Catastrophe: Some Policy Implications' *Foreign Affairs*, Winter 1983/84, p. 258.
3 William D. Casey, "A Run Worth Making," *Science*, Vol. 222, No. 4630 (December 23, 1983), p. 1281.
4 Paul Boyer, *By the Bomb's Early Light* (New York, 1985), p. 30.

5 Ibid., p. 31.
6 Ibid., p. 50.
7 Ibid.
8 Ibid., pp. 62–63.
9 Ibid., p. 60.
10 Ibid., p. 99.
11 Ibid., p. 274.
12 Paul J. Crutzen and John W. Birks, "The Atmosphere after a Nuclear War: Twilight at Noon," in Jeannie Peterson and Don Hinrichsen (eds.), *Nuclear War: The Aftermath* (Oxford, 1982), p. 84.
13 Ibid., p. 90.
14 Ibid., p. 91.
15 Paul Ehrlich, Carl Sagan, Donald Kennedy, and Walter Orr Roberts, *The Cold and the Dark: The World after Nuclear War* (New York and London, 1984), p. xiv.
16 Ronald Reagan, "Address to the Nation on Defense and National Security," March 23, 1983, http://www.reagan.utexas.edu/archives/speeches/1983/32383d.htm (accessed April 24, 2015).
17 Parade.com, "About Us," http://parade.com/about-us/ (accessed April 24, 2015).
18 Paul Ehrlich, Carl Sagan, Donald Kennedy, and Walter Orr Roberts, *The Cold and the Dark: The World after Nuclear War* (New York and London, 1984), p. xxi.
19 Philip Shabecoff, "Grimmer View Is Given of Nuclear War Effects," *New York Times*, October 31, 1983.
20 Paul Ehrlich, Carl Sagan, Donald Kennedy, and Walter Orr Roberts, *The Cold and the Dark: The World after Nuclear War* (New York and London, 1984), p. xx.
21 Carl Sagan, "Nuclear War and Climatic Catastrophe: Some Policy Implications" *Foreign Affairs*, Winter 1983/84, pp. 275 and 292.
22 R. P. Turco, O. B. Toon, T. P. Ackerman, J. B. Pollack, and Carl Sagan, "Nuclear Winter: Global Consequences of Multiple Nuclear Explosions," *Science* Vol. 222, No. 4630 (December 23, 1983), p. 1290.
23 Paul Ehrlich et al., "Long-term Biological Consequences of Nuclear War," *Science* Vol. 222, No. 4630 (December 23, 1983), p. 1299.
24 Ibid.
25 Ibid.
26 William D. Carey, "A Run Worth Making," *Science* Vol. 222, No. 4630 (December 23, 1983), p. 1281.
27 Carl Sagan, "Nuclear War and Climatic Catastrophe: Some Policy Implications" *Foreign Affairs*, Winter 1983/84, p. 292.
28 Russell Seitz, Comment and correspondence, *Foreign Affairs*, Vol. 62, No. 4 (Spring 1984), pp. 998–999.

29 Naomi Oreskes and Erik M. Conway, *Merchants of Doubt* (London, 2011), p. 65.
30 Starley L. Thompson and Stephen H. Schneider, "Nuclear Winter Reappraised," *Foreign Affairs,* Vol. 64, No. 5 (Summer 1986), p. 983.
31 Ibid., p. 998.
32 Ibid., p. 999.
33 S. Fred Singer, "Re-Analysis of the Nuclear Winter Phenomenon," *Meteorology and Atmospheric Physics* 38 (1988), pp. 228–239.
34 Starley L. Thompson and Stephen H. Schneider, "Nuclear Winter Reappraised," *Foreign Affairs,* Vol. 64, No. 5 (Summer 1986), p. 1005.
35 Ibid., p. 1002.
36 Ibid., p. 1003.
37 John Maddox, "What Happened to Nuclear Winter?" *Nature*, Vol. 333 (May 19, 1988), p. 203.
38 Ibid.
39 Henry Kissinger, *Diplomacy* (New York, 1995 edition), p. 764.
40 Ibid., p. 777.
41 Pete Earley, *Comrade J: The Untold Secrets of Russia's Master Spy in America after the End of the Cold War* (New York, 2007), back cover.
42 Ibid., p. 169.
43 Ibid., p. 171.
44 Ibid., pp. 171–172.
45 Russell Seitz, "The Melting of 'Nuclear Winter,'" *Wall Street Journal*, November 5, 1986.
46 MediaWatch, March 1994, http://archive.mrc.org/mediawatch/1994/watch19940301.asp (accessed October 7, 2016).
47 John Maddox, "What Happened to Nuclear Winter?" *Nature*, Vol. 333 (May 19, 1988), p. 203.
48 Rupert Darwall, *The Age of Global Warming* (London, 2013), p. 65.
49 UN, *The Relationship between Disarmament and Development* (New York, 1982), para. 391.
50 Ibid., para. 392.

11. Sweden Warms the World

1 http://www.scopenvironment.org/downloadpubs/scope29/statement.html (accessed July 25, 2015).
2 World Commission on Environment and Development, *Our Common Future* (Oxford and New York, 1987), p. 176.
3 Tom Selander, "Party Leaders' Views over the Next 25 Years," *Svenska Dagbladet*, November 27, 1974.
4 https://www.youtube.com/watch?v=Do9dmKuopsk, from 4:24 to 4:48 (accessed July 27, 2015).
5 Roland Huntford, *The New Totalitarians* (London, 1971), p. 152.

6 Tom Selander, "Party Leaders' Views over the Next 25 Years," *Svenska Dagbladet*, November 27, 1974.
7 David J. C. MacKay, *Sustainable Energy—with the Hot Air* (Cambridge, 2009), Fig. 24.1.
8 Gerard Caprio, "The Swedish Economy in the 1970s: The Lessons of Accommodative Policies," Federal Reserve Board International Finance Discussion Papers, No. 205, April 1982, http://www.federalreserve.gov/pubs/ifdp/1982/205/ifdp205.pdf (accessed July 26, 2015), pp. 26–27.
9 Bert Bolin, *Energi och klimat* (Stockholm, 1975), p. 1.
10 Swedish Government proposition 1975/76: No. 30 cited in Bert Bolin, *A History of the Science and Politics of Climate Change: The Role of the Intergovernmental Panel on Climate Change* (Cambridge, 2007), p. 33.
11 SAP, *Programme of the Swedish Social Democratic Party Adopted by the 1975 Congress*, p. 18.
12 *OKG-Aktuellt*, April 29, 1975.
13 Ibid., March, 1978.
14 SAP, *Social Democratic Energy Policy: Report from the Energy Policy Working Group* (Borås, 1978), p. 12.
15 Olof Palme, speech in Ronneby, February 14, 1980, http://www.olofpalme.org/wp-content/dokument/800214a_energi.pdf (accessed in July 27, 2015), p. 3.
16 Gro Harlem Brundtland, *Madam Prime Minister: A Life in Power and Politics* (New York, 2002), p. 192.
17 Ibid., p. 192.
18 Bert Bolin, *A History of the Science and Politics of Climate Change: The Role of the Intergovernmental Panel on Climate Change* (Cambridge, 2007), p. 31.
19 Ibid., p. 35.
20 Ibid., p. 36.
21 John W. Zillman, "Climate Science and Public Policy: Some Observations from Early Years at the Science-Policy Interface," Unpublished manuscript, p. 12.
22 Bert Bolin, *A History of the Science and Politics of Climate Change: The Role of the Intergovernmental Panel on Climate Change* (Cambridge, 2007), p. 38 and http://www.scopenvironment.org/downloadpubs/scope29/statement.html (accessed July 25, 2015).
23 John Maunder, "Climate Change and Villach: What is the Connection?" April 24, 2012, http://www.sunlive.co.nz/blogs/2909-climate-change-and-villach-what-is-connection.html (accessed August 7th, 2015).
24 Met Office Hadley Centre, "HadCRUT4 Global Annual Decadally Smoothed Series," http://www.metoffice.gov.uk/hadobs/hadcrut4/data/current/time_series/HadCRUT.4.4.0.0.annual_ns_avg_smooth.txt (accessed August 7th, 2015).

25 Naomi Oreskes and Erik Conway, *Merchants of Doubt: How a Handful of Scientists Obscured the Truth on Issues from Tobacco Smoke to Global Warming* (New York, Berlin, London, 2010), p. 76.
26 Royal Swedish Academy of Sciences, "The Legacy of Gordon Goodman," http://www.kva.se/globalassets/kalendarium/2015/140314-kva_legacygg_folder_090424_final.pdf (accessed August 7, 2015).
27 Bert Bolin, *A History of the Science and Politics of Climate Change: The Role of the Intergovernmental Panel on Climate Change* (Cambridge, 2007), p. 39.
28 Ibid., p. 44.
29 John W. Zillman, "Climate Science and Public Policy: Some Observations from Early Years at the Science-Policy Interface," Unpublished manuscript, p. 12.
30 Bert Bolin, *A History of the Science and Politics of Climate Change: The Role of the Intergovernmental Panel on Climate Change* (Cambridge, 2007), p. 39. Emphasis in the original.
31 John W. Zillman, "Some Observations of the IPCC Assessment Process 1988–2007," *Energy and Environment*, Vol. 18, No. 7–8, p. 871.
32 Bert Bolin, *A History of the Science and Politics of Climate Change: The Role of the Intergovernmental Panel on Climate Change* (Cambridge, 2007), p. 49.
33 WMO/UNEP, *Developing Policies for Responding to Climatic Change*, April 1988, pp. 22–23.
34 Ibid., p. 24.
35 IEA, *CO_2 Emissions from Fuel Combustion Higlights* (Paris, 2014), Table 1; Met Office Hadley Centre, "HadCRUT4 Global Annual Decadally Smoothed Series," http://www.metoffice.gov.uk/hadobs/hadcrut4/data/current/time_series/HadCRUT.4.4.0.0.annual_ns_avg_smooth.txt (accessed August 7, 2015).
36 WMO/UNEP, *Developing Policies for Responding to Climatic Change*, April 1988, p. 27.
37 Ibid., p. v.
38 WMO, "The Changing Atmosphere Implications for Global Security," Toronto, Canada, June 27–30, 1988, p. 8.
39 Ibid., p. 1.
40 The incident is captured in Palme's final interview from 9'12", https://youtu.be/Yfe_j5V84LU (accessed May 24, 2016).
41 Bo J. Theutenberg, *Dagbok Från UD 1962–1976*, Vol. 2 (Skara, 2014), p. 86.
42 https://www.youtube.com/watch?v=D09dmKuopsk from 4:48 (accessed July 27, 2015).
43 David Thomas, "The Cracks in the Greenhouse Theory," *Financial Times*, November 3/4, 1990.

44 Center for Science, Technology and Media, "Experts Express Uncertainty over Global Warming," February 13, 1992.
45 William Kininmonth email to author, July 26, 2015.
46 Ibid.

12. Sun Worship

1 Quoted from Alexander Wendt, *Der Grüne Blackout; Warum die Energiewende Nicht Functionieren Kann* (The Green Blackout: Why the Energiewende Cannot Work) (Munich, 2014), p. 17.
2 Mariah Blake, "Interview: Hermann Scheer, German Pioneer in Renewables Energies," *Christian Science Monitor*, August 20, 2008, http://www.csmonitor.com/Technology/Energy/2008/0820/interview-hermann-scheer-german-pioneer-in-renewable-energies (accessed August 8, 2015).
3 Paul Berman, *Power and the Idealists* (New York and London, 2007), p. 61.
4 WMO, "The Changing Atmosphere Implications for Global Security," Toronto, Canada, June 27–30, 1988, p. 11.
5 Energy Working Group Summary Report, June 1988, unpublished.
6 *The Economist*, "A Tax to Keep Cool," May 13, 1989.
7 Rupert Darwall, *The Age of Global Warming: A History* (London, 2013), pp. 166–167.
8 Joschka Fischer, *Der Umbau der Industriegesellschaft* (Frankfurt, 1989), pp. 55–56.
9 Ibid., pp. 81–82.
10 Ibid., p. 61.
11 Paul Berman, *Power and the Idealists* (New York and London, 2007), p. 31.
12 Anja Dalgaard-Nielsen, *Germany, Pacifism and Peace Enforcement* (Manchester and New York, 2006), p. 75.
13 Paul Berman, *Power and the Idealists* (New York and London, 2007), p. 61.
14 See Chapter 1 of Alexander Wendt, *Der Grüne Blackout; Warum die Energiewende Nicht Functionieren Kann* (The Green Blackout: Why the Energiewende Cannot Work) (Munich, 2014), p. 17.
15 Chris Turner, "The Wind at His Back," *Toronto Globe and Mail*, August 2, 2008.
16 Hermann Scheer, *A Solar Manifesto* (1993, 2nd edition, 2001, London), p. 4.
17 Margot Roosevelt, "Solar Crusader," *Time*, September 2, 2002.
18 Chris Turner, "The Wind at His Back," *Toronto Globe and Mail*, August 2, 2008.
19 Hermann Scheer, *Die Befreiung von der Bombe* (Liberation from the Bomb), (Cologne, 1986), p. 235.
20 Ibid., pp. 290–298.
21 Fred Pearce, "Bring on the Solar Revolution," *New Scientist*, May 21, 2008.
22 Hermann Scheer, *A Solar Manifesto* (London, 1993; 2nd ed. 2001), p. 4.

23 Ibid., p. 107.
24 Ibid., p. 4.
25 Ibid., p. 7.
26 Ibid., p. 8.
27 Ibid., pp. 9–10.
28 Fritz Vahrenholt interview with author, June 11, 2015.
29 Hermann Scheer, *The Energy Imperative: 100 Per Cent Renewable Now* (London and New York, 2012), p. 24.
30 Forschungsgruppe Wahlen, "Politbarometer Juli I," July 8, 2005, http://www.forschungsgruppe.de/Umfragen/Politbarometer/Archiv/Politbarometer_2005/Juli_I/ (accessed September 8, 2015).
31 *The Economist*, "The Statecraft of Davela Merkeron," August 29, 2015.
32 Fritz Vahrenholt interview with author, June 11, 2015.
33 The Department of Trade and Industry paper was leaked to *The Guardian* and can be accessed via Ashley Seager and Mark Milner, "Revealed: Cover-Up Plan on Energy Target," theguardian.com (August 13, 2007), http://www.theguardian.com/environment/2007/aug/13/renewable energy.energy.
34 William J. Antholis, "The Good, the Bad, and the Ugly: EU-U.S. Cooperation on Climate Change," June 10, 2009, http://www.brookings.edu/research/speeches/2009/06/10-climate-antholis.
35 Derived from illustrative annual yield of 1 kW solar PV array using standard polycrystalline modules fixed at optimal tilt, National Renewable Energy Laboratory, PV Watts Calculator.
36 Friedrich Nietzsche (tr. R. J. Hollingdale), *Thus Spoke Zarathustra* (London, 1969), p. 39.
37 Hermann Scheer, *A Solar Manifesto* (1993, 2nd edition, 2001, London), p. 8.
38 F. A. Hayek, *The Road to Serfdom* (1944; Abingdon and New York, reprinted 2006), pp. 172–173.
39 Ibid., p. 178.
40 Ralf Georg Reuth, *Goebbels: The Life of Joseph Goebbels, the Mephistophelean Genius of Nazi Propaganda* (London, 1995), p. 41.
41 Sean O'Hagan, "Man of Mystery" *The Observer*, January 30, 2005, http://www.theguardian.com/artanddesign/2005/jan/30/art2 (accessed September 8, 2015); and Ulrike Knöfel, "Beuys Biography: Book Accuses Artist of Close Ties to Nazis," Spiegelonline, May 17, 2013, http://www.spiegel.de/international/germany/new-joseph-beuys-biography-discloses-ties-to-nazis-a-900509.html (accessed September 9, 2015).

13. Renewable Destruction

1 Federal Ministry for the Environment, Nature Conservation, Building and Nuclear Safety, "Erneuerbare-Energien-Gesetz tritt in Kraft" ("Renewable Energy Law Comes into Force"), July 30, 2004, http://www.bmub.bund

.de/presse/pressemitteilungen/pm/artikel/erneuerbare-energien-gesetz-tritt-in-kraft/ (accessed August 12, 2015).

2 Reuters, "German 'Green Revolution' May Cost 1 Trillion Euros—Minister," February 20, 2013, http://www.reuters.com/article/2013/02/20/us-germany-energy-idUSBRE91J0AV20130220 (accessed September 2, 2015).

3 Quoted from Alexander Wendt, *Der Grüne Blackout; Warum die Energiewende Nicht Functionieren Kann* (The Green Blackout: Why the Energiewende Cannot Work) (Munich, 2014), p. 11.

4 Alexander Wendt, *Der Grüne Blackout; Warum die Energiewende Nicht Functionieren Kann* (The Green Blackout: Why the Energiewende Cannot Work) (Munich, 2014), p. 15.

5 Hans Poser, Jeffrey Altman, Felix ab Egg, Andreas Granata, and Ross Board, *Development and Integration of Renewable Energy: Lessons Learned from Germany* (July 2014), Figure 16, www.finadvice.ch.

6 *The Economist*, "The Cost del sol," July 20, 2013, http://www.economist.com/news/business/21582018-sustainable-energy-meets-unsustainable-costs-cost-del-sol (accessed September 25, 2015).

7 Competition and Markets Authority, *Energy Market Investigation: Provisional Findings Report* (July 2015), para. 2.144.

8 Ibid., para. 5.180.

9 Rupert Darwall, *Central Planning with Market Features: How Renewable Subsidies Destroyed the UK Electricity Market* (London, 2015), p. 44.

10 Competition and Markets Authority, *Energy Market Investigation: Provisional Findings Report* (July 2015), para. 7.7.

11 Office for Budget Responsibility, *Economic and Fiscal Outlook*, Cm 9088 (July 2015), Table 4.5.

12 Competition and Markets Authority, *Energy Market Investigation: Provisional Findings Report* (July 2015), para. 2.74.

13 Ibid., para. 2.80.

14 Deutscher Bundestag, Unterrichtung durch die Bundesregierung: Gutachten zu Forschung, Innovation und technologischer Leistungsfähigkeit Deutschlands 2014 (March 2014), p. 49; and Fritz Vahrenholt interview with author, June 6, 2015.

15 Alexander Wendt, *Der Grüne Blackout; Warum die Energiewende Nicht Functionieren Kann* (The Green Blackout: Why the Energiewende Cannot Work) (Munich, 2014), p. 16.

16 Ibid., p. 19.

17 Ibid., p. 20.

18 Deutscher Bundestag, Unterrichtung durch die Bundesregierung: Gutachten zu Forschung, Innovation und technologischer Leistungsfähigkeit Deutschlands 2014 (March 2014), p. 49.

19 Ibid.

20 Manuel Frondel, Nolan Ritter, Christoph M. Schmidt, and Colin Vance,

"Economic Impacts from the Promotion of Renewable Energy Technologies: The German Experience" (2010), http://www.rwi-essen.de/media/content/pages/publikationen/ruhr-economic-papers/REP_09_156.pdf, pp. 15–16.

21 Ibid.

22 Fritz Vahrenholt, "Energiewende and Climate Change—from the German Vanguard Role to a Fiasco" (unpublished, 2015).

23 Bundesamt für Umwelt (Federal Office for the Environment), "Energieverbrauch der privaten Haushalte," February 19, 2015, https://www.umweltbundesamt.de/daten/private-haushalte-konsum/energieverbrauch-der-privaten-haushalte.

24 Nuclear Energy Institute, "Fukushima, Chernobyl and the Nuclear Event Scale," Summer 2011, http://www.nei.org/News-Media/News/News-Archives/fukushima-chernobyl-and-the-nuclear-event-scale (accessed September 27, 2015).

25 Spiegel Online International, "Aftermath of Election Debacle: Merkel Fires Environment Minister Röttgen," May 16, 2012, http://www.spiegel.de/international/germany/chancellor-angela-merkel-sacks-environment-minister-norbert-roettgen-a-833614.html.

26 Spiegel Online International, "Germany's Energy Poverty: How Electricity Became a Luxury Good," September 4, 2013, http://www.spiegel.de/international/germany/high-costs-and-errors-of-german-transition-to-renewable-energy-a-920288.html.

27 Fritz Vahrenholt interview with author, June 6, 2015.

28 Alexander Wendt, *Der Grüne Blackout; Warum die Energiewende Nicht Functionieren Kann* (The Green Blackout: Why the Energiewende Cannot Work) (Munich, 2014), p. 58.

29 James Kanter and Jad Mouawad, "Money and Lobbyists Hurt European Efforts to Curb Gases," *New York Times*, December 10, 2008, http://www.nytimes.com/2008/12/11/business/worldbusiness/11carbon.html.

30 Index and share price data taken from www.finanzen.net and company website (WE).

31 BDEW Bundesverband der Energie- und Wasserwirtschaft (Federal Association of Energy and Water Industries), "Strompreisanalyse März 2015," https://www.bdew.de/internet.nsf/id/9D1CF269C1282487C1257E22002BC8DD/$file/150409%20BDEW%20zum%20Strompreis%20der%20Haushalte%20Anhang.pdf.

32 Estimated on the basis of prorating the loss in respect of Vattenfall.

33 Steve Johnson, "Private Equity Retreats from Renewables 'Fad,'" *Financial Times*, February 16, 2014, http://www.ft.com/cms/s/0/ef1b2248-94bb-11e3-9146-00144feab7de.html#axzz3nDFwoqIp.

34 Company annual reports.

35 Klaus Peter Krause, "Immer am Rand eines grossen Stromausfalls" ("Always on the edge of a major power outage"), EIKE, September 24, 2014,

http://www.eike-klima-energie.eu/news-cache/immer-am-rand-eines-grossen-stromausfalls/ (accessed September 29, 2015).

36 Catalina Schröder, "Energy Revolution Hiccups: Grid Instability Has Industry Scrambling for Solutions," Spiegel Online International, August 16, 2012, http://www.spiegel.de/international/germany/instability-in-power-grid-comes-at-high-cost-for-german-industry-a-850419.html (accessed September 29, 2015).

37 Manuel Frondel, Nolan Ritter, Christoph M. Schmidt, and Colin Vance, "Economic Impacts from the Promotion of Renewable Energy Technologies: The German Experience," November 2009, http://www.rwi-essen.de/media/content/pages/publikationen/ruhr-economic-papers/REP_09_156.pdf, p. 15.

38 Ibid., p. 14.

39 Federal Ministry of Economics and Technology, "German Mittelstand: Engine of the German Economy," (undated), https://www.bmwi.de/English/Redaktion/Pdf/factbook-german-mittelstand,property=pdf,bereich=bmwi2012,sprache=en,rwb=true.pdf, p. 3.

40 BDEW Bundesverband der Energie- und Wasserwirtschaft e.V. (Federal Association of Energy and Water Industries), Strompreisanalyse März 2015, https://www.bdew.de/internet.nsf/id/9D1CF269C1282487C1257E22002BC8DD/$file/150409%20BDEW%20zum%20Strompreis%20der%20Haushalte%20Anhang.pdf.

41 Eric Heymann, "Carbon Leakage: A Barely Perceptible Process," Deutsche Bank Research, January 23, 2014, p. 1.

42 Ibid., p. 9.

43 Ibid., p. 8.

44 Ibid., p. 1.

45 Eric Heymann, "Capital Investment in Germany at Sectoral Level," Deutsche Bank Research, January 9, 2015, p. 7.

46 Ibid.

47 Ibid., p. 1.

48 Rupert Darwall, *Central Planning with Market Features: How Renewable Subsidies Destroyed the UK Electricity Market* (London, 2015), pp. 14–15.

49 Bill Peacock and Josiah Neeley, "The Cost of the Production Tax Credit and Renewable Energy Subsidies in Texas," Texas Public Policy Foundation, November 2012.

50 Vladimir Rakov and Martin Uman, *Lightning: Physics and Effects* (Cambridge, 2003), Table 1.

51 Mark Mills, "The Clean Power Plan Will Collide with the Incredibly Weird Physics of the Electric Grid," forbes.com, August 7, 2015, http://www.forbes.com/sites/markpmills/2015/08/07/the-clean-power-plan-will-collide-with-the-incredibly-weird-physics-of-the-electric-grid/ (accessed October 2, 2015).

14. The Curse of Intermittency

1 National Academy of Engineering Press Release, "National Academy of Engineering Reveals Top Engineering Impacts of the 20th Century: Electrification Cited as Most Important," February 22, 2000.
2 Ludwig von Mises, *A Critique of Interventionism* (1929, rev. 1976, tr. 1996), p. 9.
3 http://www.c-span.org/video/?155533-1/engineering-20th-century (accessed July 20, 2016)
4 National Academy of Engineering Press Release, "National Academy of Engineering Reveals Top Engineering Impacts of the 20th Century: Electrification Cited as Most Important," February 22, 2000.
5 http://www.c-span.org/video/?155533-1/engineering-20th-century (accessed July 20, 2016).
6 Lion Hirth, *The Economics of Wind and Solar Variability: How the Variability of Wind and Solar Power Affects Their Marginal Value, Optimal Deployment, and Integration Costs* (Berlin, 2014), p. 36.
7 Ibid., p. 40.
8 Ibid., p. 39.
9 U.S. Energy Information Administration, U.S. Gulf Coast Conventional Gasoline Regular Spot Price FOB (dollars per gallon) updated to September 16, 2015, http://tonto.eia.gov/dnav/pet/hist/LeafHandler.ashx?n=PET&s=EER_EPMRU_PF4_RGC_DPG&f=D.
10 Paul L. Joskow, "Comparing the Cost of Intermittent and Dispatchable Electricity Generating Technologies," September 27, 2010 (rev. February 9, 2011), p. 1.
11 Lion Hirth, *The Economics of Wind and Solar Variability: How the Variability of Wind and Solar Power Affects Their Marginal Value, Optimal Deployment, and Integration Costs* (Berlin, 2014), p. 40.
12 Paul L. Joskow, "Comparing the Cost of Intermittent and Dispatchable Electricity Generating Technologies," September 27, 2010 (rev. February 9, 2011), p. 1.
13 Ibid., p. 22.
14 Tom Steyer, "Tom Steyer: 50% Clean Energy by 2030," EcoWatch.com, July 24, 2015, http://ecowatch.com/2015/07/24/steyer-clean-energy-plan/ (accessed October 6, 2015).
15 For example, Apple Inc.'s claim on its environmental responsibility corporate webpage that 100 percent of its U.S. operations and all its data centers run on renewable energy, http://www.apple.com/environment/ (accessed on October 6, 2015).
16 Stephen Lacey "Google Engineers Explain Why They Stopped R&D in Renewable Energy," greentechmedia November 19, 2014, http://www.greentechmedia.com/articles/read/google-engineers-explain-why-they-stopped-rd-in-renewable-energy (accessed on December 8, 2015).

17 Christopher Adams and John Thornhill, "Gates to Double Investment in Renewable Energy Projects," ft.com, June 25, 2015, http://www.ft.com/cms/s/2/4f66ff5c-1a47-11e5-a130-2e7db721f996.html#axzz3eejgP31M (accessed October 6, 2015).
18 M. J. Grubb, "Value of Variable Sources on Power Systems," *IEE Proceedings—C*, Vol. 138, No. 2 (March 1991), p. 151.
19 *The Daily Telegraph*, "Era of Constant Electricity at Home Is Ending, Says Power Chief," March 2, 2011.
20 Hermann Scheer, *A Solar Manifesto* (London, 1993; 2nd ed., 2001), p. 118.
21 Ibid., p. 120.
22 OECD/NEA, *Nuclear Energy and Renewables: System Effects in Low-carbon Electricity Systems* (2012), p. 37.
23 Lion Hirth, *The Economics of Wind and Solar Variability: How the Variability of Wind and Solar Power Affects Their Marginal Value, Optimal Deployment, and Integration Costs* (Berlin, 2014), p. 87.
24 Ibid., p. 88.
25 Ibid., p. 94.
26 Ibid., p.
27 Ibid., p. 119.
28 Ibid., p. 118.
29 Ibid., p. 183.
30 OECD/NEA, *Nuclear Energy and Renewables: System Effects in Low-carbon Electricity Systems* (2012), p. 37.
31 Ibid., p. 33.
32 Rupert Darwall, "Britain's Energy Market Is Back under State Control," capx.co, April 8, 2015, http://www.capx.co/britains-energy-market-is-back-under-state-control/ (accessed October 7, 2015).
33 C. Frank, *The Net Benefits of Low and No-Carbon Electricity* (Washington, D.C., 2014), Table 9A.
34 Siemens, "One Year after Fukushima—Germany's Path to a New Energy Policy," March 2012, http://www.siemens.com/press/pool/de/feature/2012/corporate/2012-03-energiewende/factsheet-e.pdf (accessed October 7, 2015).
35 e.on, "No Economic Prospects: Owners of the Irsching 4 and 5 Gas-fired Power Stations Announce Their Closure," March 30, 2015, http://www.eon.com/en/media/news/press-releases/2015/3/30/no-economic-prospects-owners-of-the-irsching-4-and-5-gas-fired-power-stations-announce-their-closure.html.
36 Alexander Wendt, *Der Grüne Blackout; Warum die Energiewende Nicht Functionieren Kann* (The Green Blackout: Why the Energiewende Cannot Work) (Munich, 2014), p. 49.
37 Fritz Vahrenholt interview with author, June 11, 2015.
38 Thomas Friedman, "Germany, the Green Superpower," *New York Times*, May 6, 2014.

39 Paul Berman, *Power and the Idealists* (New York and London, 2007), p. 309.
40 Thomas Friedman, "Germany, the Green Superpower," *New York Times*, May 6, 2014.

15. Climate Industrial Complex

1 Robert Cooke, *Dr. Folkman's War: Angiogenesis and the Struggle to Defeat Cancer* (New York, 2001), p. 99.
2 Ibid., p. 78.
3 Ibid., pp. 81–82.
4 Günter Hartkopf, "Umweltverwaltung—eine organisatorische Herausforderung' (Environmental Management—an Organizational Challenge), Bad Kissingen, January 8, 1986.
5 Paul Hockenos, *Joshka Fischer and the Making of the Berlin Republic: An Alternative History of Postwar Germany* (Oxford, 2008), p. 254.
6 Dagmar Dehmer, "Bundesrechnungshof bezweifelt Sinn der Deutschen Energieagentur," *Der Tagesspiel*, November 17, 2014, http://www.tagesspiegel.de/politik/bilanztricks-bundesrechnungshof-bezweifelt-sinn-der-deutschen-energieagentur/10991908.html (accessed October 22, 2015).
7 Fritz Vahrenholt interview with author, June 11, 2015.
8 James Kanter, "Scientist: Warming Could Cut Population to 1 Billion," nytimes.com, March 13, 2009, http://dotearth.blogs.nytimes.com/2009/03/13/scientist-warming-could-cut-population-to-1-billion/?_r=0 (accessed October 22, 2015).
9 David Krutzler, "Die Natur wird deutlich zu uns sprechen," *Der Standard*, October 9, 2013, http://derstandard.at/1379293635977/Die-Natur-wird-deutlich-zu-uns-sprechen?ref=article (accessed October 22, 2015).
10 Fides News Agency, "Press Conference for the Presentation of the Encyclical Laudato Sì," June 18, 2015, http://www.news.va/en/news/press-conference-for-the-presentation-of-the-encyc (accessed October 22, 2015).
11 Pope Francis, *Encyclical Letter* Laudato Sì *of the Holy Father on Care for Our Common Home* (Rome, May 24, 2105), p. 21.
12 Bernhard Pötter, "Klimapolitik verteilt das Weltvermögen neu' (Climate policy to redistribute global wealth), *Neue Zürcher Zeitung*, November 14, 2010.
13 MISEREOR, http://www.misereor.org/en/about-us/lenten-campaign.html (accessed October 22, 2015).
14 Richard Read, "Solar CEO Frank Asbeck Sees United States as "God's Country" for Green Power," *The Oregonian*, October 27, 2010, http://www.oregonlive.com/business/index.ssf/2010/10/larworld_ceo_frank_asbeck_sees.html (accessed October 22, 2015).
15 Ibid.
16 UPI, "Ailing German PV Panel Maker SolarWorld Completes Restructur-

ing," February 26, 2014, http://www.upi.com/Science_News/Technology/2014/02/26/Ailing-German-PV-panel-maker-SolarWorld-completes-restructuring/2582139339098o/ (accessed October 22, 1015).

17 Share price data from reuters.com and market capitalization and other financial data from SolarWorld AG annual reports.

18 Michael Grubb Linked-in page, https://uk.linkedin.com/pub/michael-grubb/1b/581/78a; Climate Strategies about webpage, http://climatestrategies.org/about-us/; The Carbon Trust about us webpage, http://www.carbontrust.com/about-us/ (all accessed October 23, 2015).

19 Climate Strategies website search term "funders," http://climatestrategies.org/?s=funders (accessed October 23, 2015).

20 Bent Flyvbjerg, "Survival of the Unfittest: Why the Worst Infrastructure Gets Built—and What We Can Do about It," *Oxford Review of Economic Policy*, Vol. 25, No. 3 (2009), p. 345.

21 Ibid., pp. 350–351.

22 Ibid., p. 358.

23 Ibid.

24 Ibid., p. 349.

25 Fritz Vahrenholt interview with author, June 11, 2015.

26 Günter Hartkopf, "Umweltverwaltung—eine organisatorische Herausforderung' (Environmental Management—an Organizational Challenge), Bad Kissingen, January 8, 1986.

27 Fred Pearce, *Green Warriors: The People and the Politics behind the Environmental Revolution* (London, 1991), p. 283.

28 Ibid., p. 285.

29 Ibid., p. 287.

16. Power without Responsibility

1 Clive Hambler and Susan Canney, *Conservation* (2nd ed., Cambridge, 2013), p. 1.

2 Jeremy Leggett, *The Carbon War: Global Warming and the End of the Oil Era* (New York, 2001), p. 193.

3 Ibid., p. vii.

4 Fred Pearce, *Green Warriors: The People and the Politics behind the Environmental Revolution* (London, 1991), p. 18.

5 Patrick Moore, "Should We Celebrate Carbon Dioxide?" 2015 Annual GWPF Lecture, October 15, 2015, http://www.thegwpf.org/patrick-moore-should-we-celebrate-carbon-dioxide/.

6 Fred Pearce, *Green Warriors: The People and the Politics behind the Environmental Revolution* (London, 1991), p. 41.

7 Greenpeace International, "Golden Rice," http://www.greenpeace.org/international/en/campaigns/agriculture/problem/Greenpeace-and-Golden-Rice/ (accessed October 28, 2015).

8 Jeremy Leggett, *The Carbon War: Global Warming and the End of the Oil Era* (New York, 2001), p. 100.
9 Cintia Cheong, "Munich Re Takes Number One Spot in Top 15 Global Reinsurers Rankings," *The Actuary*, July 15, 2015, http://www.theactuary.com/news/2015/07/munich-re-takes-number-one-spot-in-top-15-global-reinsurers-rankings/#sthash.G9scFcR9.dpuf (accessed October 27, 2015).
10 Matthew J. Belvedere, "No Climate Change Impact on Insurance Biz: Buffett," CNBC, March 3, 2014, http://www.cnbc.com/2014/03/03/no-climate-change-impact-on-insurance-biz-buffett.html (accessed October 27, 2015).
11 John Dizard, "Sunshine and Blue Skies Create Climate Tragedy for Reinsurers," *Financial Times*, January 26, 2015.
12 Jeremy Leggett, *The Carbon War: Global Warming and the End of the Oil Era* (New York, 2001), p. 101.
13 Jeremey Leggett, "Climate Change and the Financial Sector," *Journal of the Society of Fellows*, Vol. 9, No. 2 (January 1995), p. 119.
14 Ibid., pp. 121–122.
15 Ibid., p. 129.
16 Ibid., p. 132.
17 Ibid., p. 135.
18 Ibid., p. 136.
19 Jeremy Leggett, *The Carbon War: Global Warming and the End of the Oil Era* (New York, 2001), p. 244.
20 Solarplaza, "UK Solar Market Awaits Major Boost," March 29, 2010, http://www.prnewswire.co.uk/news-releases/uk-solar-market-awaits-major-boost-152896315.html (accessed October 28, 2015).
21 Solar Century Holdings Limited Report and Financial statements for 2015, p. 2.
22 Ibid., p. 24.
23 https://twitter.com/jeremyleggett/status/406691588950073344 (accessed October 28, 2015).
24 Tessa Kipping, "Jeremy Leggett: It's Flattering When People Dismiss Solar Power," *BusinessGreen*, December 16, 2013, http://www.businessgreen.com/bg/interview/2318947/jeremy-leggett-its-flattering-when-people-dismiss-solar-power (accessed October 28, 2015).
25 Florian Diekmann, "Kostenexplosion bei Strom, Öl, Gas: Energiearmut in Deutschland nimmt drastisch zu," ("Cost explosion in electricity, oil, gas: energy poverty in Germany increases dramatically"), Spiegelonline, February 24, 2014, http://www.spiegel.de/wirtschaft/service/gruenen-anfrage-energiearmut-in-deutschland-nimmt-drastisch-zu-a-954688.html (accessed October 28, 2015).
26 Spiegelonline, "Germany's Energy Poverty: How Electricity Became a Luxury Good," April 9, 2013, http://www.spiegel.de/international/

germany/high-costs-and-errors-of-german-transition-to-renewable-energy-a-920288.html (accessed October 28, 2015).

27 Alexander Wendt, *Der Grüne Blackout; Warum die Energiewende Nicht Functionieren Kann* (The Green Blackout: Why the Energiewende Cannot Work) (Munich, 2014), p. 4.

28 Alexander Wendt email to author, November 6, 2015.

29 Seb Beloe, John Elkington, Katie Fry Hester, and Sue Newell, *The 21st Century NGO: In the Market for Change* (London, 2003), p. 2.

30 Richard Edelman, "Rebuilding Public Trust through Accountability and Responsibility: Address to the Ethical Corporation Magazine Conference," New York, October 3, 2002, Slide 12.

31 Ibid., Slide 11.

32 Robert Mendick and Edward Malnick, "European Union Funding £90m Green Lobbying Con," www.telegraph.co.uk, December 21, 2013, http://www.telegraph.co.uk/news/worldnews/europe/10532853/European-Union-funding-90m-green-lobbying-con.html (accessed October 28, 2015).

33 Green 10, http://www.green10.org/about-us/ (accessed October 28, 2015).

34 Green 10, http://www.green10.org/publications/ (accessed October 28, 2015).

35 Financial data from EU Transparency Register accessed via link for each NGO at bottom, http://www.green10.org/about-us/ (accessed October 28, 2015).

36 Emily Gosden, "Greenpeace Executive Flies 250 Miles to Work," www.telegraph.co.uk, June 23, 2014, http://www.telegraph.co.uk/news/earth/earthnews/10920198/Greenpeace-executive-flies-250-miles-to-work.html.

37 WWF, "We Must Save the World's Wild Life: An International Declaration," (The Morges Manifesto) April 29, 1961 accessed via http://wwf.panda.org/who_we_are/history/sixties/ (accessed November 2, 2015).

38 Clive Hambler, "Wind Farms vs. Wildlife," *The Spectator*, January 5, 2013.

39 Rachel Carson, *Silent Spring* (London, 2000), p. 22.

40 Ronald Bailey, *The End of Doom: Environmental Renewal in the Twenty-first Century* (New York, 2015), pp. 111–112.

41 For example, Wendy Koch, "Wind Turbines Kill Fewer Birds Than Do Cats, Cell Towers," *USA Today*, September 15, 2014, http://www.usatoday.com/story/money/business/2014/09/15/wind-turbines-kill-fewer-birds-than-cell-towers-cats/15683843/.

42 J. R. Settele, R. Scholes, R. Betts, S. Bunn, P. Leadley, D. Nepstad, J. T. Overpeck, and M. A. Taboada, 2014, "Terrestrial and Inland Water Systems." In: *Climate Change 2014: Impacts, Adaptation, and Vulnerability. Contribution of Working Group II to the Fifth Assessment Report of the Intergovernmental Panel on Climate Change* (Cambridge and New York), p. 327.

43 Clive Hambler, "Where Eagles Dare—the Wind Farms Gamble," *Wild Land News*, No. 83 (June 2013), p. 10.
44 Clive Hambler, "Wind Farms vs. Wildlife," *The Spectator*, January 5, 2013.
45 Clive Hambler and Susan Canney, *Conservation* (2nd ed., Cambridge, 2013), pp. 58–59.
46 David Rose, "Where HAVE all our woods gone? Up in smoke—as the new trendy 'green' wood-burning stoves and boilers (funded by tax millions) are being fuelled by birches and oaks . . . leaving swathes of Britain barren," *Mail on Sunday*, June 14, 2015.
47 David Rose, "The UK's £1 billion carbon-belcher raping US forests . . . that YOU pay for: How world's biggest green power plant is actually INCREASING greenhouse gas emissions and Britain's energy bill," *Mail on Sunday*, June 7, 2015.
48 Alexis Schwarzenbach, *Saving the World's Wildlife: WWF—the First 50 Years* (London, 2011), p. 136.
49 Ibid., pp. 136–137.
50 Ibid., p. 164.
51 Ibid., pp. 169–170.
52 Ibid., p. 165.
53 Fred Pearce, *Green Warriors: The People and the Politics behind the Environmental Revolution* (London, 1991), p. 15.
54 Alexis Schwarzenbach, *Saving the World's Wildlife: WWF—the First 50 Years* (London, 2011), p. 276.
55 John Elkington with Tom Burke, *The Green Capitalists: Industry's Search for Environmental Excellence* (London, 1987), p. 221.
56 Fred Pearce, *Green Warriors: The People and the Politics behind the Environmental Revolution* (London, 1991), p. 60.

17. Swallowing Hard

1 Fred Pearce, *Green Warriors: The People and the Politics behind the Environmental Revolution* (London, 1991), p. 15.
2 http://fortune.com/global500/volkswagen-8/ (accessed November 15, 2015).
3 http://www.volkswagen-nabu.de/ueber_die_kooperation/meilensteine.
4 NABU, "How Far VW is on the Road to Ecological No. 1?" December 2, 2013, https://www.nabu.de/news/2013/12/16428.html (accessed November 3, 2015).
5 Volkswagen AG, "Volkswagen und NABU gründen internationalen Moorschutzfonds," March 11, 2015, http://www.volkswagenag.com/content/vwcorp/info_center/de/news/2015/03/fonds.html; and taz.de "Treuer Partner von VW," http://taz.de/Umweltverband-steht-zu-Autokonzern/!5232353/ (both accessed November 3, 2015).
6 Max von Biederback, Alexander Neubacher, and Gerald Traufetter, "Geld oder Klage," *Der Spiegel*, March 25, 2103.

7 Maximilian Weingartner, “Die seltsame Allianz von Greenpeace und Volkswagen” (“The strange alliance between Greenpeace and Volkswagen”), *Frankfurter Allgemeine Zeitung*, December 1, 2015.
8 NABU and Volkswagen, “The Cooperation,” No. 7, April 2014, http://www.volkswagen-nabu.de/download/07%2014-05-19_Koop_VW_NABU_EN_klein.pdf (accessed November 3, 2015).
9 EPA press notice, “EPA, California Notify Volkswagen of Clean Air Act Violations/Carmaker Allegedly Used Software That Circumvents Emissions Testing for Certain Air Pollutants,” September 18, 2015.
10 Michael Cames and Eckard Helmers, “Critical Evaluation of the European Diesel Car Boom—Global Comparison, Environmental Effects and Various National Strategies, *Environmental Sciences Europe* 2013, 25:15, Fig. 2.
11 See, for example, Chapter 5.3.1.2 “Improving Drive Train Efficiency’ of IPCC, *Climate Change 2007: Working Group III: Mitigation of Climate Change*, https://www.ipcc.ch/publications_and_data/ar4/wg3/en/ch5s5-3-1-2.html (accessed November 4, 2015).
12 Michael Cames and Eckard Helmers, “Critical Evaluation of the European Diesel Car Boom—Global Comparison, Environmental Effects and Various National Strategies,” *Environmental Sciences Europe* 2013, 25:15, p. 5.
13 Ibid., p. 9.
14 Ibid., p. 10.
15 Ibid., pp. 10–11.
16 Ibid., p. 10.
17 Quality of Urban Air Report Group, *Diesel Vehicle Emissions and Urban Air Quality* (December 1993), p. ii.
18 Dominic Lawson, “Take a Deep Breath and Say It, Brussels: The US Is Right about Pollution,” *The Sunday Times*, September 27, 2015.
19 European Commission, “Communication from the Commission to the European Parliament, the Council, the European Economic and Social Committee and the Committee of the Regions: A Policy Framework for Climate and Energy in the Period from 2020 to 2030,” COM (2014) 15 final, January 22, 2014, p. 9.
20 Steffen Skovmand, “Europe’s Renewables Reversal,” *Wall Street Journal*, February 4, 2014.
21 Alexander Wendt email to author, November 1, 2015.
22 Manuel Frondel, Nolan Ritter, Christoph M. Schmidt, and Colin Vance, “Economic Impacts from the Promotion of Renewable Energy Technologies: The German Experience,” November 2009, http://www.rwi-essen.de/media/content/pages/publikationen/ruhr-economic-papers/REP_09_156.pdf, p. 19.
23 Robert Wilson, “Why Germany’s Nuclear Phase-Out Is Leading to More Coal Burning,” Carbon Counter blog, June 6, 2015, https://carboncounter.wordpress.com/2015/06/06/why-germanys-nuclear-phaseout-is-leading-to-more-coal-burning/ (accessed November 11, 2015).
24 Alexander Wendt, *Der Grüne Blackout; Warum die Energiewende Nicht*

Functionieren Kann (The Green Blackout: Why the Energiewende Cannot Work) (Munich, 2014), p. 30.

25 D. S. Nussbaum and S. Reinis, "Some Individual Differences in Human Response to Infrasound," UTIAS Report No. 282, January 1984.

26 Clive Hambler and Susan Canney, *Conservation* (2nd ed., Cambridge, 2013), p. 60.

27 Ministry of Ecology, Energy and Sustainable Development, *Energy Transition for Green Growth Act* (October 2015), p. 3.

28 Energy Information Administration, *August 2015 Monthly Energy Review*, Table 12.6.

18. Golden to Green

1 Arne Næss (tr. and ed. David Rothenberg), *Ecology, Community and Lifestyle* (Cambridge, 1989), p. 185.

2 Barbara Wood, *Alias Papa: A Life of Fritz Schumacher* (London, 1984), p. 248.

3 Joel Kotkin, *The New Class Conflict* (Candor, NY, 2014), p. 8.

4 Steven Malanga, "The Green behind California's Greens," *City Journal*, Spring 2015.

5 Joel Kotkin, *The New Class Conflict* (Candor, NY, 2014), p. 11.

6 Joel Kotkin, "Are We Heading for an Economic Civil War?," *The Daily Beast*, September 8, 2015, http://www.thedailybeast.com/articles/2015/11/08/are-we-heading-for-an-economic-civil-war.html.

7 Joel Kotkin, *The New Class Conflict* (Candor, NY, 2014), p. 94.

8 Ibid.

9 Ibid., p. 85.

10 Ibid., p. 65.

11 Joel Kotkin, "Are We Heading for an Economic Civil War?" *The Daily Beast*, September 8, 2015, http://www.thedailybeast.com/articles/2015/11/08/are-we-heading-for-an-economic-civil-war.html.

12 Matt Rosoff, "Eric Schmidt: We Don't Talk about Occupy Wall Street in the Valley Because We Don't Have Those Problems," SFChronicle.com, December 23, 2011, http://www.sfgate.com/news/article/ERIC-SCHMIDT-We-Don-t-Talk-About-Occupy-Wall-2424084.php.

13 Pacific Gas and Electric Company Press Release, "Pacific Gas and Electric Company Files for Chapter 11 Reorganization," April 6, 2001.

14 California Energy Commission, "Energy Almanac," Total System Power, http://energyalmanac.ca.gov/electricity/total_system_power.html; and Total Production by Resource Type, http://energyalmanac.ca.gov/electricity/electricity_generation.html (accessed January 10, 2016).

15 Arnold Schwarzenegger, Executive Order S-3-05, June 1, 2005.

16 Edmund G. Brown Jr., "Inaugural Address Remarks as Prepared," January 5, 2015, https://www.gov.ca.gov/news.php?id=18828 (accessed January 13, 2016).

17 Michael Kanellos, "Arnold Lashes Out at Valero, Prop 23," Greentech

media.com, September 27, 2010, http://www.greentechmedia.com/articles/read/arnold-lashes-out-at-valero-prop-23-opponents (accessed January 13, 2016).

18 John Seiler, "Do High Taxes Cost State Jobs?" CalWatchdog.com, August 30, 2010, http://calwatchdog.com/2010/08/30/do-high-taxes-cost-state-jobs/ (accessed January 13, 2016).

19 Ballotpedia, "California Proposition 23, the Suspension of AB 32 (2010), https://ballotpedia.org/California_Proposition_23,_the_Suspension_of_AB_32_(2010)#cite_ref-36 (accessed January 13, 2016).

20 Joel Kotkin, *The New Class Conflict* (Candor, NY, 2014), p. 38.

21 Petra Bartosiewicz and Marissa Miley, *The Too Polite Revolution: Why the Recent Campaign to Pass Comprehensive Climate Legislation in the United States Failed* (January 2013), p. 29.

22 Jessica Brown, Jeff Cole, Melinda Hanson, Sivan Kartha, Michael Lazarus, Andreas Merkl, Heather Thompson, Mitch Tobin, and Laura Wolfson, *Design to Win: Philanthropy's Role in the Fight against Global Warming* (San Francisco, 2007), p. 12.

23 Petra Bartosiewicz and Marissa Miley, *The Too Polite Revolution: Why the Recent Campaign to Pass Comprehensive Climate Legislation in the United States Failed* (January 2013), pp. 29–30.

24 https://twitter.com/ClimateWorks (accessed January 14, 2016).

25 Petra Bartosiewicz and Marissa Miley, *The Too Polite Revolution: Why the Recent Campaign to Pass Comprehensive Climate Legislation in the United States Failed* (January 2013), p. 31.

26 Ibid., p. 8.

27 Jessica Brown, Jeff Cole, Melinda Hanson, Sivan Kartha, Michael Lazarus, Andreas Merkl, Heather Thompson, Mitch Tobin, and Laura Wolfson, *Design to Win: Philanthropy's Role in the Fight against Global Warming* (San Francisco, 2007), p. 12.

28 U.S. Senate Committee on Environment and Public Works (Minority), *The Chain of Environmental Command: How a Club of Billionaires and Their Foundations Control the Environmental Movement and Obama's EPA* (July 30, 2014), pp. 6–7.

29 Ibid., p. 22.

30 Ibid., p. 13.

31 K. Shawn Smallwood, "Comparing Bird and Bat Fatality-Rate Estimates among North American Wind-Energy Projects," *Wildlife Society Bulletin*, Vol. 37, No. 1 (2013).

32 John Flicker, "Audubon Statement on Wind Power," *Audubon Magazine*, November–December 2006, retrieved via http://www.millbrookwindfarm.ca/uploads/files/audubon-statement-wind-power.pdf (accessed January 15, 2016).

33 Carl Levesque, "For the Birds: Audubon Society Stands Up in Support of Wind Energy," renewablenergyworld.com, December 14, 2006, http://

www.renewableenergyworld.com/articles/2006/12/for-the-birds-audubon-society-stands-up-in-support-of-wind-energy-46840.html.

34 U.S. Senate Committee on Environment and Public Works (Minority), *The Chain of Environmental Command: How a Club of Billionaires and Their Foundations Control the Environmental Movement and Obama's EPA* (July 30, 2014), p. 22.

35 Ballotpedia, "California Proposition 23, the Suspension of AB 32 (2010), https://ballotpedia.org/California_Proposition_23,_the_Suspension_of_AB_32_(2010)#cite_ref-36 (accessed January 13, 2015).

36 Rebecca A. Kagan, Tabitha C. Viner, Pepper W. Trail and Edgar Espinoza, "Avian Mortality at Solar Energy Facilities in Southern California: A Preliminary Analysis," April 2014, pp. 1–2, accessed via http://alternativeenergy.procon.org/sourcefiles/avian-mortality-solar-energy-ivanpah-apr-2014.pdf (accessed January 15, 2016).

37 Ellen Knickmeyer and John Locher, "'Alarming" Rate of Bird Deaths as New Solar Plants Scorch Animals in Mid-Air," October 18, 2014, http://www.huffingtonpost.com/2014/08/18/bird-deaths-solar_n_5686700.html (accessed January 15, 2016).

38 American Bird Conservancy, "Bald and Golden Eagles Victorious: Court Invalidates 30-Year 'Eagle Take' Rule," August 12, 2015, http://abcbirds.org/article/bald-golden-eagles-victorious-court-invalidates-30-year-eagle-take-rule/ (accessed January 15, 2016).

39 American Bird Conservancy, http://abcbirds.org/program/wind-energy/ (accessed January 15, 2016).

40 National Audubon Society, https://www.audubon.org/content/audubons-position-wind-power (accessed January 15, 2016).

19. Capitalism's Fort Sumter

1 Adam Smith, *An Inquiry into the Nature and Causes of the Wealth of Nations* (Chicago, 1976), p. 363.

2 Mark Mills interview with author, June 27, 2015.

3 Mark Mills, "Will the Paris Climate Summit Lead to More Money for Scientists or Solyndras?" Forbes.com, December 2, 2015, http://www.forbes.com/sites/markpmills/2015/12/02/progress-in-paris-at-the-cop21-more-money-for-scientists-or-corporatists/print/ (accessed January 15, 2016).

4 Steven Malanga, "The Green behind California's Greens," *City Journal*, Spring 2015.

5 Adam Smith, *An Inquiry into the Nature and Causes of the Wealth of Nations* (Chicago, 1976), p. 477.

6 Ibid., p. 478.

7 Joel Kotkin, "Are We Heading for an Economic Civil War?" *The Daily Beast*, September 8, 2015, http://www.thedailybeast.com/articles/2015/11/08/are-we-heading-for-an-economic-civil-war.html.

8 U.S. Senate Committee on Environment and Public Works (Minority),

The Chain of Environmental Command: How a Club of Billionaires and Their Foundations Control the Environmental Movement and Obama's EPA (July 30, 2014), p. 13.

9 Vivian Krause, "Rockefellers behind 'Scruffy Little Outfit,'" *Financial Post*, February 14, 2013, http://business.financialpost.com/fp-comment/rockefellers-behind-scruffy-little-outfit?_federated=1.

10 U.S. Senate Committee on Environment and Public Works (Minority), *The Chain of Environmental Command: How a Club of Billionaires and Their Foundations Control the Environmental Movement and Obama's EPA* (July 30, 2014), p. 8.

11 Jarol Manheim, *Biz-War and the Out-of-Power Elite: The Progressive-Left Attack on the Corporation* (Mahwah, NJ, 2004), p. 52.

12 Ibid., p. 41.

13 Ibid., p. 82.

14 The Pew Charitable Trusts, Consolidated Financial Statements and Report of Independent Certified Public Accountants, June 30, 2014 and 2013, p. 5.

15 Jarol Manheim, *Biz-War and the Out-of-Power Elite: The Progressive-Left Attack on the Corporation* (Mahwah, NJ, 2004), p. 126.

16 Ibid., p. 126.

17 Martin Morse Wooster, "The Inscrutable Billionaire," *Philanthropy*, Summer 2008.

18 Foundation Center, http://foundationcenter.org/findfunders/topfunders/top100assets.html (accessed January 18, 2016); and U.S. Senate Committee on Environment and Public Works (Minority), *The Chain of Environmental Command: How a Club of Billionaires and Their Foundations Control the Environmental Movement and Obama's EPA* (July 30, 2014), p. 13.

19 *Science*, "Possible Acid Rain Woes in the West," Vol. 228, No. 4695 (April 5, 1985), p. 34.

20 Richard North, "Global Governance: Funding the NGO Monster," EU Referendum.com, July 7, 2014, http://www.eureferendum.com/blogview.aspx?blogno=85066 (accessed January 18, 2016).

21 Joel Kotkin, *The New Class Conflict* (Candor, NY, 2014), p. 39.

22 Steven Malanga, "The Green Behind California's Greens," *City Journal*, Spring 2015.

23 Kathleen Hartnett-White and Stephen Moore, *Fueling Freedom: Exposing the Mad War on Energy* (Washington, D.C., 2016) p. 48.

24 Pope Francis, Address to the FAO, Rome June 11, 2015, http://www.fao.org/about/meetings/conference/c2015/address-pope-francis/en/.

25 LIUNA, "LIUNA Leaves BlueGreen Alliance," January 20, 2012, http://www.liuna.org/news/story/liuna-leaves-bluegreen-alliance (accessed January 19, 2016).

26 *Wall Street Journal*, "California's Climate Change Revolt," September 11,

2015, http://www.wsj.com/articles/californias-climate-change-revolt-1442014369 (accessed September 14, 2015).

20. The Washington, D.C., *Energiewende*

1 George W. Bush, State of the Union address, January 23, 2007.
2 Barack Obama, "Remarks by the President on America's Energy Security, Georgetown University Washington, D.C.," March 30, 2011, https://www.whitehouse.gov/the-press-office/2011/03/30/remarks-president-americas-energy-security.
3 George W. Bush, State of the Union address, January 23, 2007.
4 George W. Bush, State of the Union address, January 28, 2008.
5 U.S. Energy Information Administration, U.S. Natural Gas Marketed Production, https://www.eia.gov/dnav/ng/hist/n9050us2a.htm (accessed January 26, 2016).
6 Mark P. Mills, *Where The Jobs Are: Small Businesses Unleash America's Energy Employment Boom* (Manhattan Institute, February 2014), p. 3.
7 Barack Obama, State of the Union address, January 25, 2011.
8 U.S. Energy Information Administration, U.S. Natural Gas Marketed Production, https://www.eia.gov/dnav/ng/hist/n9050us2a.htm; and U.S. Field Production of Crude Oil, https://www.eia.gov/dnav/pet/hist/LeafHandler.ashx?n=PET&s=MCRFPUS2&f=A (accessed January 26, 2016).
9 Barack Obama, State of the Union address, January 24, 2012.
10 Amanda Cooper, "IEA Says Oil Market May 'Drown in Oversupply' in 2016," Reuters, January 19, 2016, http://www.reuters.com/article/us-oil-iea-idUSKCN0UX0VJ (accessed January 27, 2016)
11 Barack Obama, State of the Union address, January 24, 2012.
12 Mark P. Mills, *Where the Jobs Are: Small Businesses Unleash America's Energy Employment Boom* (Manhattan Institute, February 2014), pp. 3–4.
13 Ibid., pp. 3 and 10.
14 U.S. Senate Committee on Environment and Public Works (Minority), *The Chain of Environmental Command: How a Club of Billionaires and Their Foundations Control the Environmental Movement and Obama's EPA* (July 30, 2014), p. 39.
15 Media Impact Funders, http://mediaimpactfunders.org/festival/gasland/ (accessed January 28, 2016).
16 Adelaide Park Gomer, "New York Should Become the First State to Ban Fracking," Alternet, December 6, 2011, http://www.alternet.org/story/153336/new_york_should_become_the_first_state_to_ban_fracking (accessed January 28, 2016).
17 Ibid.
18 Thomas Kaplan, "Citing Health Risks, Cuomo Bans Fracking in New York State," *New York Times*, December 17, 2014, http://www.nytimes.com/

2014/12/18/nyregion/cuomo-to-ban-fracking-in-new-york-state-citing-health-risks.html?_r=1 (accessed January 28, 2016).

19 U.S. Senate Committee on Environment and Public Works (Minority), *The Chain of Environmental Command: How a Club of Billionaires and Their Foundations Control the Environmental Movement and Obama's EPA* (July 30, 2014), p. 22.

20 U.S. Senate Committee on Environment and Public Works (Majority), *Obama's Carbon Mandate: An Account of Collusion, Cutting Corners and Costing Americans Billions* (August 4, 2015), p. 16.

21 Ibid., p. 15.

22 Foxnews.com, "Administration Warns of 'Command-and-Control' Regulation over Emissions," December 9, 2009, http://www.foxnews.com/politics/2009/12/09/administration-warns-command-control-regulation-emissions.html (accessed January 29, 2016).

23 The White House, "Press Conference by the President," November 3, 2010, https://www.whitehouse.gov/the-press-office/2010/11/03/press-conference-president (accessed January 29, 2016).

24 U.S. Senate Committee on Environment and Public Works (Majority), *Obama's Carbon Mandate: An Account of Collusion, Cutting Corners and Costing Americans Billions* (August 4, 2015), p. 7.

25 Ibid., p. 5.

26 Ibid., p. 18.

27 Ibid.

28 U.S. Senate Committee on Environment and Public Works (Minority), *The Chain of Environmental Command: How a Club of Billionaires and Their Foundations Control the Environmental Movement and Obama's EPA* (July 30, 2014), pp. 25–29.

29 Darren Samuelsohn, "'NRDC mafia' finding homes on Hill, in EPA," *New York Times*, March 6, 2009, http://www.nytimes.com/gwire/2009/03/06/06greenwire-nrdc-mafia-finding-homes-on-hill-in-epa-10024.html (accessed February 1, 2016).

30 C. J. Ciaramella, "Ex-EPA Administrator Lisa Jackson Contacted Lobbyist from Private Email," *The Washington Free Beacon*, August 14, 2013, http://www.washingtontimes.com/news/2013/aug/14/ex-epa-administrator-lisa-jackson-contacted-lobbyi/?page=all (accessed February 1, 2016).

31 Supreme Court of the United States, *Utility Air Regulatory Group v. Environmental Protection Agency*, 573 U.S. (2014), p. 14.

32 Supreme Court of the United States, *Massachusetts v. Environmental Protection Agency*, 549 U.S. (2007), p. 30.

33 Ibid., p. 20.

34 Supreme Court of the United States, *Massachusetts v. Environmental Protection Agency*, 549 U.S. (2007), Roberts, C. J., Dissenting, p. 8.

35 *Federal Register*, Vol. 74, No. 239 (December 15, 2009), pp. 66497–66498.

36 U.S. Senate Committee on Environment and Public Works (Majority), *Obama's Carbon Mandate: An Account of Collusion, Cutting Corners and Costing Americans Billions* (August 4, 2015), p. 29.
37 U.S. Senate Committee on Environment and Public Works (Minority), *The Chain of Environmental Command: How a Club of Billionaires and Their Foundations Control the Environmental Movement and Obama's EPA* (July 30, 2014), p. 22.
38 James Bennett, "Pandering for Profit: This Transformation of Health Charities to Lobbyists," George Mason Department of Economics Paper No. 11-54, (December 14, 2011), pp. 3–4.
39 ALA, "American Lung Association Hails Bipartisan Rejection of Clean Air Rollback," November 10, 2011, http://ala1-old.pub30.convio.net/press-room/press-releases/bipartisan-rejection-rollback.html (accessed February 2, 2011).
40 James Bennett, "Pandering for Profit: This Transformation of Health Charities to Lobbyists," George Mason Department of Economics Paper No. 11-54, (December 14, 2011), p. 22.
41 The White House, "President Obama Announces Historic 54.5 mpg Fuel Efficiency Standard," July 29, 2011, https://www.whitehouse.gov/the-press-office/2011/07/29/president-obama-announces-historic-545-mpg-fuel-efficiency-standard (accessed February 2, 2016).
42 The White House, "Obama Administration Finalizes Historic 54.5 MPG Fuel Efficiency Standards," August 28, 2012, https://www.whitehouse.gov/the-press-office/2012/08/28/obama-administration-finalizes-historic-545-mpg-fuel-efficiency-standard (accessed February 2, 2016).
43 EPA, *Regulatory Impact Analysis: Final Rulemaking for 2017–2025 Light-Duty Vehicle Greenhouse Gas Emission Standards and Corporate Average Fuel Economy Standards* (August 1, 2012), p. 3–18 and Table 2; and Energy Information Administration, "U.S. Regular All Formulations Retail Gasoline Prices (Dollars per Gallon)," http://tonto.eia.gov/dnav/pet/hist/LeafHandler.ashx?n=PET&s=EMM_EPMR_PTE_NUS_DPG&f=A (accessed December 20, 2015).
44 On the basis annual consumption is three percent higher than 2014 (i.e., 140.9b gallons) https://www.eia.gov/forecasts/steo/report/us_oil.cfm.
45 Peter W. Huber and Mark P. Mills, *The Bottomless Well: The Twilight of Fuel, the Virtue of Waste, and Why We Will Never Run out of Energy*, (New York, 2006), p. 105.
46 Mark P. Mills, *Where the Jobs Are: Small Businesses Unleash America's Energy Employment Boom* (Manhattan Institute, February 2014), p. 3.
47 U.S. Senate Committee on Environment and Public Works (Majority), *Obama's Carbon Mandate: An Account of Collusion, Cutting Corners and Costing Americans Billions* (August 4, 2015), p. 33.
48 Barack Obama, State of the Union address, February 12, 2013.

21. "Our Kids' Health"

1 Gina McCarthy, testimony before the Senate Committee on Environment and Public Works, July 23, 2014.
2 The White House, "Remarks by the President in Announcing the Clean Power Plan," August 3, 2105, https://www.whitehouse.gov/the-press-office/2015/08/03/remarks-president-announcing-clean-power-plan (accessed February 17, 2016).
3 Coral Davenport and Gardiner Harris, "Obama to Unveil Tougher Environmental Plan with His Legacy in Mind," *New York Times*, August 2, 2015, http://www.nytimes.com/2015/08/02/us/obama-to-unveil-tougher-climate-plan-with-his-legacy-in-mind.html?_r=1 (accessed March 1, 2016).
4 The White House, "Remarks by the President in Announcing the Clean Power Plan," August 3, 2105, https://www.whitehouse.gov/the-press-office/2015/08/03/remarks-president-announcing-clean-power-plan (accessed February 17, 2016).
5 EPA, *Regulatory Impact Analysis for the Clean Power Plan Final Rule* (October 23, 2015), Tables 3–21 and 3–8; Chris Bryant, "Soaring Renewable Energy Costs Set to Stoke German Energy Debate," ft.com, October 15, 2013, http://www.ft.com/cms/s/0/03486cb0-3587-11e3-b539-00144feab7de.html#axzz41wVy2YkG (accessed March 4, 2016); electricity consumption (2014): U.S.—3,903 TWh, Germany—614 TWh, respectively, U.S. Energy Information Administration, *February 2016 Monthly Energy Review*; and Bundesministerium für Wirtschaft und Energie, Statistisches Bundesamt, Arbeitsgruppe Erneuerbare Energien-Statistik (AGEE-Stat), "Zahlen und Fakten Energiedaten," Stromerzeugungskapazitäten, Bruttostromerzeugung und Bruttostromverbrauch Deutschland, Tab 22, http://www.bmwi.de/DE/Themen/Energie/energiedaten.html.
6 Derived from Tables 8–1 and 8–2, EPA, *Regulatory Impact Analysis for the Clean Power Plan Final Rule* (October 23, 2015).
7 EPA, *Regulatory Impact Analysis for the Clean Power Plan Final Rule* (October 23, 2015), pp. 3–28.
8 Ibid., Table 3-9.
9 EPA, Technical Support Document (TSD) for Carbon Pollution Guidelines for Existing Power Plants: Emission Guidelines for Greenhouse Gas Emissions from Existing Stationary Sources: Electric Utility Generating Units, Greenhouse Gas Mitigation Measures, p. 4-2, Table 4-1.
10 Gina McCarthy, testimony before the Senate Committee on Environment and Public Works, July 23, 2014.
11 EPA, Technical Support Document (TSD) for Carbon Pollution Guidelines for Existing Power Plants: Emission Guidelines for Greenhouse Gas Emissions from Existing Stationary Sources: Electric Utility Generating Units, Greenhouse Gas Mitigation Measures, p. 4-2, Table 4-1.
12 Average acre/MW (5 MW/KM^2) from NREL, U.S. Renewable Energy Tech-

nical Potentials: A GIS-Based Analysis, July 2012; state areas from U.S. Census, Geography, State Area Measurements.

13 EPA, Technical Support Document (TSD) for Carbon Pollution Guidelines for Existing Power Plants: Emission Guidelines for Greenhouse Gas Emissions from Existing Stationary Sources: Electric Utility Generating Units, Greenhouse Gas Mitigation Measures, p. 4-2, Table 4-1; and EIA, International Energy Statistics, Renewables, 2012.

14 Ibid.

15 EPA, *Regulatory Impact Analysis for the Clean Power Plan Final Rule* (October 23, 2015), Table 3–9.

16 Ibid., Table 3–3.

17 UNFCCC, "U.S. Cover Note, INDC and Accompanying Information," submitted on March 31, 2015 accessed via http://www4.unfccc.int/submissions/indc/Submission%20Pages/submissions.aspx (accessed March 8, 2016).

18 U.S. Energy Information Administration, "February 2016 Monthly Energy Review' (February 24, 2016), Table 7.1 Electricity Overview.

19 Peter W. Huber and Mark P. Mills, *The Bottomless Well: The Twilight of Fuel, the Virtue of Waste, and Why We Will Never Run out of Energy* (New York, 2006), p. 107.

20 Ibid., pp. 94–95.

21 Ibid., p. 57.

22 Emphasis in the original. Peter W. Huber and Mark P. Mills, *The Bottomless Well: The Twilight of Fuel, the Virtue of Waste, and Why We Will Never Run out of Energy* (New York, 2006), p. 48.

23 Jennifer Oldham, "As Pot-Growing Expands, Electricity Demands Tax U.S. Grids," Bloomberg.com, December 21, 2015, http://www.bloomberg.com/news/articles/2015-12-21/as-pot-growing-expands-power-demands-tax-u-s-electricity-grids (accessed March 8, 2016).

24 Interagency Working Group on Social Cost of Carbon, USG, *Technical Support Document:—Technical Update of the Social Cost of Carbon for Regulatory Impact Analysis—Under Executive Order 12866—*, May 2013, revised July 2015, p. 14.

25 Derived from Chris Hope, *The Social Cost of Carbon from the PAGE09 model* (June 2011), Table 9.

26 Ibid., p. 18.

27 EPA, *Regulatory Impact Analysis for the Clean Power Plan Final Rule* (October 23, 2015), pp. 4–5.

28 William Pizer et al., "Using and Improving the Social Cost of Carbon," *Science*, Vol. 346, No. 6214, p. 1190.

29 US Department of State, "Congressional Budget Justification: Foreign Assistance Fiscal Year 2015," Table 1.

30 EPA, *Regulatory Impact Analysis for the Clean Power Plan Final Rule* (October 23, 2015), Tables 8-1 and 8-2.

31 William A. Knudson, "The Environment, Energy, and the Tinbergen Rule," *Bulletin of Science Technology and Society*, Vol. 29, No. 4 (August 2009), pp. 308–312.
32 William M. Hodan and William R. Barnard, *Evaluating the Contribution of $PM_{2.5}$ Precursor Gases and Re-entrained Road Emissions to Mobile Source $PM_{2.5}$ Particulate Matter Emissions* (undated), p. 3, https://www3.epa.gov/ttnchie1/conference/ei13/mobile/hodan.pdf.
33 David J. Nowak, Satoshi Hirabayashi, Allison Bodine, Robert Hoehn, "Modelled $PM_{2.5}$ Removal by Trees in Ten US Cities and Associated Health Effects," *Environmental Pollution*, 178 (2013), p. 398.
34 EPA, *Regulatory Impact Analysis for the Clean Power Plan Final Rule* (October 23, 12015), pp. 4–21.
35 Kathleen Hartnett-White, *The EPA's Pretense of Science: Regulating Phantom Risks* (Texas Public Policy Foundation, May 2012), p. 12.
36 EPA, *Summary of Expert Opinions on the Existence of a Threshold Concentration-Response Function for $PM_{2.5}$-related Mortality* (June 2010), p. 16.
37 Dr. Ann Smith quoted in Kathleen Hartnett-White, *The EPA's Pretense of Science: Regulating Phantom Risks* (Texas Public Policy Foundation, May 2012), p. 12.
38 Ross McKitrick email to author, March 18, 2016.

22. Saving the Planet

1 Pascal Bruckner (tr. Steven Rendall), *The Fanaticism of the Apocalypse* (Cambridge and Malden, MA, 2013), p. 29.
2 Ibid., pp. 56–57.
3 Ibid., p. 178.
4 Jeffrey Goldberg, "The Obama Doctrine," *The Atlantic*, April 2016, http://www.theatlantic.com/magazine/archive/2016/04/the-obama-doctrine/471525/ (accessed March 14, 2016).
5 UNFCCC, "U.S. Cover Note, INDC and Accompanying Information," submitted on March 31, 2015, accessed via http://www4.unfccc.int/submissions/indc/Submission%20Pages/submissions.aspx (March 8, 2016).
6 Article Four (1) (b).
7 The White House, "U.S. Leadership and the Historic Paris Agreement to Combat Climate Change," December 12, 2015, https://www.whitehouse.gov/the-press-office/2015/12/12/us-leadership-and-historic-paris-agreement-combat-climate-change (accessed March 15, 2016).
8 AP, "Climate Deal Must Avoid U.S. Congress Approval, French Minister Says," June 1, 2015 via http://www.theguardian.com/world/2015/jun/01/un-climate-talks-deal-us-congress (accessed March 15, 2016).
9 Daniel Bodansky, "In Brief: Legal Options for U.S. Acceptance of a New Climate Change Agreement," C2ES, May 2015, p. 4.

10 Steven Groves, "Obama's Plan to Avoid Senate Review of the Paris Protocol," The Heritage Foundation, September 21, 2015, pp. 4–5 via http://www.heritage.org/research/reports/2015/09/obamas-plan-to-avoid-senate-review-of-the-paris-protocol.
11 Daniel Bodansky, "Legally Binding Versus Non-legally Binding Instruments," in Scott Barrett, Carlo Carraro and Jaime de Melo, *Towards a Workable and Effective Climate Regime* (London and Clermont–Ferrand, 2015), p. 159.
12 The White House, "Press Conference by President Obama," December 1, 2015, https://www.whitehouse.gov/the-press-office/2015/12/01/press-conference-president-obama (accessed March 15, 2016).
13 Donald B. Verrilli Jr., "On Application for Immediate Stay of Final Agency Action: Memorandum for the Federal Respondents in Opposition," February 2016, pp. 71–72.
14 Reuters, "U.S. Top Court Rules against Obama Administration over Air Pollution Rule," June 29, 2015, http://www.reuters.com/article/usa-court-pollution-idUSL2N0ZF11Y20150629 (accessed March 15, 2016).
15 Supreme Court of the United States, *Utility Air Regulatory Group v. EPA*, 573 U.S., p. 2.
16 The White House, "Statement by the Press Secretary," February 9, 2016, https://www.whitehouse.gov/the-press-office/2016/02/10/statement-press-secretary (accessed March 16, 2016).
17 Barbara Lewis, "U.S. Will Sign Paris Agreement and Stick to It—Stern," Reuters, February 16, 2016, http://www.reuters.com/article/us-climate change-summit-usa-idUSKCN0VP1VA.
18 EPA, "Administrator Gina McCarthy, Remarks on Climate Action at CERA in Houston Texas, as Prepared," February 24, 2016, https://yosemite.epa.gov/opa/admpress.nsf/8d49f7ad4bbcf4ef852573590040b7f6/5c432a7068e191e985257f630054fea8!OpenDocument (accessed March 16, 2016).
19 Robert Redford, "We Cannot Let SCOTUS Block Clean Power," *Time*, February 21, 2016, http://time.com/4231875/robert-redford-climate-change/ (accessed March 16, 2016).
20 Pascal Bruckner (tr. Steven Rendall), *The Fanaticism of the Apocalypse* (Cambridge and Malden, MA, 2013), pp. 180–181.
21 David Siders, "Jerry Brown: "Never Underestimate the Coercive Power of the Central State,'" sacbee.com, December 7, 2015, http://www.sacbee.com/news/politics-government/capitol-alert/article48466200.html (accessed March 17, 2016).
22 Jasper Scherer, "Has Sanders Pushed Clinton to the Left on Energy Issues?" *US News & World Report*, March 8, 2016, http://www.usnews.com/news/articles/2016-03-08/has-sanders-pushed-clinton-to-the-left-on-energy-issues (accessed March 17, 2016).
23 Emily Atkin, "BREAKING: Hillary Clinton Endorses Federal Investigation

of Exxon," ThinkProgress, October 29, 2015, http://thinkprogress.org/climate/2015/10/29/3717602/clinton-investigate-exxon/ (accessed March 17, 2016).

24 J. T. Houghton, G. J. Jenkins, and J. J. Ephraums, *Climate Change: The IPCC Scientific Assessment* (Cambridge, 1990), p. 254.

23. Spiral of Silence

1 John Locke, *An Essay Concerning Human Understanding*, Chapter 18, para. 12.

2 Edward Alsworth Ross, *Social Control: A Survey of the Foundations of Order* (New York, 1901), p. 105.

3 Keith N. Hampton et al., *Social Media and the "Spiral of Silence"* (Pew Research Center, Washington, D.C., August 26, 2014), p. 1.

4 Ibid., p. 18.

5 Ibid., p. 1.

6 *The Times*, "Professor Elisabeth Noelle-Neumann," April 15, 2010.

7 Quoted in Christopher Simpson, "Elisabeth Noelle-Neumann's Spiral of Silence and the Historical Context of Communication Theory." *Journal of Communication*, Vol. 46, No. 3 (September 1996), p. 149.

8 David Childs, "Elisabeth Noelle-Neumann: Pioneer of Public-opinion Polling and Market Research," *The Independent*, April 9, 2010, http://www.independent.co.uk/news/obituaries/elisabeth-noelle-neumann-pioneer-of-public-opinion-polling-and-market-research-1940766.html (accessed April 3, 2016).

9 Quoted in Denis McQuail, "Paradigm Shifts in the Study of Media Effects," in Wolfgang Donsbach, Charles T. Salmon and Yariv Tsfati (eds), *The Spiral of Silence: New Perspectives on Communication and Public Opinion* (New York and London, 2014), 135.6.

10 Elisabeth Noelle-Neumann, *The Spiral of Silence: Public Opinion—Our Social Skin* (Chicago and London, 1984), p. 6.

11 Ibid., p. ix.

12 Ibid., pp. 56–57.

13 Ibid., p. ix.

14 Ibid., p. 46.

15 Christopher Simpson, "Elisabeth Noelle-Neumann's Spiral of Silence and the Historical Context of Communication Theory," *Journal of Communication*, Vol. 46, No. 3 (September 1996), p. 156.

16 William H. Honan, "US Professor's Criticism of German Scholar's Work Stirs Controversy," *New York Times*, August 27, 1997.

17 Elisabeth Noelle-Neumann, *The Spiral of Silence: Public Opinion—Our Social Skin* (Chicago and London, 1984), p. 78.

18 James Madison, *The Federalist* No. 49, in *The Federalist Papers* (New York, 1982), p. 256.

19 Elisabeth Noelle-Neumann, *The Spiral of Silence: Public Opinion—Our Social Skin* (Chicago and London, 1984), p. 77.
20 Ibid., p. 173.
21 Ibid., p. 148. Emphasis in the original.
22 Ibid.
23 The White House, "President Barack Obama's State of the Union Address," January 28, 2014, https://www.whitehouse.gov/the-press-office/2014/01/28/president-barack-obamas-state-union-address (accessed April 5, 2016).
24 Stephen R. C. Hicks, *Explaining Postmodernism: Skepticism and Socialism from Rousseau to Foucault* (Phoenix and New Berlin, WI, 2004), p. 178.
25 John F. Kerry, "Remarks on Climate Change," February 16, 2014, http://www.state.gov/secretary/remarks/2014/02/221704.htm (accessed April 5, 2016).
26 Quoted in Michael Booth, *The Almost Nearly Perfect People: The Truth about the Nordic Miracle* (London, 2014), p. 345.
27 Michael Booth, *The Almost Nearly Perfect People: The Truth about the Nordic Miracle* (London, 2014), p. 336.
28 Ibid.
29 Ibid., pp. 349–350.

24. The American Republic

1 Charles de Secondat, Baron de Montesquieu, *De l'esprit des lois* (1748) Book 8, Chapter 1.
2 Jon Entine, "How the Columbia Journalism School Smeared Exxon," *New York Post*, March 1, 2016, http://nypost.com/2016/03/01/how-the-columbia-journalism-school-smeared-exxon/ (accessed 6 April 2016).
3 Jagadish Shukla et al., "Letter to President Obama, Attorney General Lynch, and OSTP Director Holdren," http://web.archive.org/web/20150920110942/http:/www.iges.org/letter/LetterPresidentAG.pdf; and *Wall Street Journal*, "The Climate Change 1%," March 2, 2016, http://www.wsj.com/articles/the-climate-change-1-1456886481 (accessed 6 April 2016).
4 *State of West Virginia, et al. v. United States Environmental Protection Agency et al., Brief of Amici Curiae Amazon.Com, Inc., Apple Inc., Google Inc., and Microsoft Corp. in Support of Respondents* (April 1, 2016), p. 6.
5 Ibid., p. 2.
6 Ibid., p. 10, fn. 4.
7 Peter Gay, *The Enlightenment: An Interpretation Vol. 2—The Science of Freedom* (first published 1970; London 1979), p. 552.
8 Ibid., p. 489.
9 Ibid., p. 490.

10 Alexander Hamilton, *The Federalist* No. 1, in *The Federalist Papers* (New York, 1982), p. 2.
11 Peter Gay, *The Enlightenment: An Interpretation Vol. 2—The Science of Freedom* (first published 1970; London 1979), p. 566.
12 James Madison, *The Federalist* No. 49, in *The Federalist Papers* (New York, 1982), p. 256.

Index